Applied Mineralogy
Technische Mineralogie

Edited by
Herausgegeben von

V. D. Fréchette, Alfred, N.Y.
H. Kirsch, Essen
L. B. Sand, Worcester, Mass.
F. Trojer, Leoben

5

Springer-Verlag
New York Wien 1973

D. McConnell

Apatite
Its Crystal Chemistry, Mineralogy,
Utilization, and Geologic
and Biologic Occurrences

Springer-Verlag
New York   Wien   1973

DUNCAN McCONNELL, B. S., M. S., Ph.D., Professor of Mineralogy and of Dental
Research at The Ohio State University, Columbus, Ohio, U.S.A.

REF
QE
391
.A6
M3

This work is subject to copyright.
All rights are reserved,
whether the whole or part of the material is concerned,
specifically those of translation, reprinting, re-use of illustrations,
broadcasting, reproduction by photocopying machine
or similar means, and storage in data banks.
© 1973 by Springer-Verlag/Wien
Library of Congress Catalog Card Number 72-88060
Printed in Austria

With 17 Figures

ISBN 0-387-81095-1 Springer-Verlag New York-Wien
ISBN 3-211-81095-1 Springer-Verlag Wien-New York

Preface

The preparation of a volume on this topic was undertaken with some hesitancy on my part because the ramifications of the mineralogy of apatite involve both biological and physical sciences in very elaborate ways. This hesitancy may have arisen in part from the realization that considerable skill would be required in order to extract the meaning from the thousands of papers that have appeared within the past twenty years; the task of attempting to extract and assemble the usable information seemed gigantic. Greatly adding to the difficulty was the fact that a considerable portion of these journal articles contain nothing of value and further confuse a most complex topic.

Nevertheless, it was thought that some of my formal education in the biological sciences, which has been greatly extended and augmented during the past fifteen years, might be integrated with my more extensive education and experience in chemistry, crystallography, mineralogy, geology and physics in order to produce something that would relate to the mineral apatite and its extremely diverse occurences in nature. At the same time it seemed essential to point out some of the many important aspects in which this knowledge bears on geology, agriculture, chemical engineering, medicine and dentistry.

Perhaps the most difficult question that confronted me was: What level of educational background should be expected for a potential reader? What proficiency in inorganic chemistry and/or crystallography should be expected, as examples?

Obviously one cannot supply adequate instruction on such diversified topics and, at the same time, proceed toward an orderly presentation of the principal subject. My intention, then, became one of attempting to supply a source of data and current interpretations to graduate students in both physical and biological sciences. Thus, although a textbook presentation of crystal chemistry has been avoided, it is believed that there is an urgent need for adequate explanation of the fundamental principles that are essential to some of the interpretations that have been applied.

Complex theoretical discussions are omitted—particularly on such topics as phosphors—but recent references to the literature are supplied, and it is my hope that some readers will be inspired to delve into these matters in greater depth than would be appropriate for inclusion in this small volume.

No pretense is made with regard to complete coverage of the worldwide occurrences of apatite. Indeed, many earlier analyses are not mentioned except in connection with crucial interpretations or unusual characteristics, and even in the case of recent information of critical significance, the analytical data frequently are omitted when readily accessible elsewhere. Nevertheless, examples are given—particularly for calculations that have not been presented previously. Thus, an attempt has been made to present fundamental information that is not cluttered with details of questionable significance. Of the thousands of papers that have been published—many of them in journals which certainly do not have critical reviewers—only a relatively small number have been cited in this work. In general, there was a need for selectivity among references, and two criteria were used: a fairly recent paper by an author who has demonstrated some proficiency on the particular topic, or an earlier paper that indicates some priority with respect to the development of a particular concept. Thus, there are few discussions of concepts that are incompatible with accepted theory. Numerous references have been omitted in consistency with this policy.

While some space is given to the history of development of the topic—particularly in the introduction and under the heading earlier work—this discussion has more value than may be immediately obvious insofar as it reveals that some completely disproven theories have returned to hount modern scientific interpretations. It is mildly amazing, for example, that some still refer to fluorapatite as a "double salt" (tricalcium phosphate and calcium fluoride) in the light of modern crystallochemical knowledge, but it has been my unfortunate experience to discover inorganic chemists who do so.

There is no thorough discussion of experimental methods, and perhaps this topic could be covered adequately by a few authors in a volume several times larger than this one. Some commentary is given on the general appropriateness of certain methods in connection with specific questions. Many of the conclusions bearing on the structural intricacies of carbonate apatites that are based on data obtained by infrared absorption spectroscopy are the crudest sort of guesses, but such conclusions are stated in such complex ways as to imply respectability. In some instances it is almost unbelievable that the investigators could be so unfamiliar with the capabilities and limitations of the methods.

Methodology seems to develop as fads, however, and the current variety of self-mesmerism involves vibrational spectra within a comparatively narrow range of the electromagnetic spectrum. Observations of optical properties are infrequently reported for the many synthetic concoctions, possibly because the authors are completely unfamiliar with the use of a polarizing microscope for such purposes. Probably the most deplorable situation exists with respect to chemical analysis; an analytical chemist seems to be becoming a person who supposedly knows which buttons to push and in what sequence.

While improvements in methods of analysis surely must be welcomed, it must be admitted by any thoughtful person that there is no computational device that is capable of deductive reasoning, and such devices are completely dependent upon a series of instructions which are and must be as fallible as the humans who supply these instructions. Many answers have been obtained, but some of these answers have become disassociated from the questions that were asked and therefore are meaningless.

A discerning reader will notice minor discrepancies, such as the unit-cell volume for a particular composition being given as one value in the chapter on physical properties and a slightly different value in the chapter on crystal chemistry. In general, such differences are insignificant with respect to the purposes of the calculations involved. Furthermore, there is agreement between the values given by various authors only within certain limits, so our purposes are best served by small inconsistencies, rather than being consistently in error or consistently accurate — and without knowing which.

In general, the arrangement of the topics is a progression from simpler to more complex concepts, and I wish to make a plea that the reader not attempt to read the chapter on biologic apatites without careful consideration of the earlier chapter on carbonate apatites. Although an attempt has been made to avoid repetition — through cross references — it has seemed necessary to mention some critical results in more than one place in order to preserve the continuity of a particular discourse.

Columbus, Ohio, Fall 1972   DUNCAN MCCONNELL

Acknowledgments

Initially I wish to pay tribute to a series of most competent teachers from whom I had the good fortune to receive formal instruction: M. H. STOW and JAMES LEWIS HOWE (both deceased, but formerly at Washington and Lee University), A. C. GILL and H. RIES (both deceased, but formerly at Cornell University), W. H. ZACHARIASEN, D. JEROME FISHER and A. JOHANNSEN (at the University of Chicago, the last is deceased), AUSTIN F. ROGERS (deceased, formerly at Stanford University) and J. W. GRUNER, E. B. SANDELL, R. B. ELLESTAD and FRANK F. GROUT (at the University of Minnesota, the last is deceased). To this list should be added P. M. HARRIS, at the Ohio State University; he has been a source of friendly, constructive criticism during the past twenty years.

With special reference to this work, I am indebted to many persons, particularly VAN DERCK FRÉCHETTE (College of Ceramics, State University of New York at Alfred) for editorial advice. My colleague and erstwhile graduate student, DENNIS W. FOREMAN, JR., critically read the manuscript, which was typed by Miss PAMELA DAVIS. The Dean of the College of Dentistry has encouraged this project by making suitable facilities available. Grants from the U.S. Public Health Service (both the Surgeon General's Office and the National Institute of Dental Research) and a contract with the U.S. Army Medical Research and Development Command made possible much of the experimental work during the past fifteen years. However, my gratitude also must be expressed to the Procter & Gamble Company, the Edward Orton, Jr. Ceramic Foundation, and the University Development Fund [O.S.U.].

The following persons have contributed useful information, in several cases beyond what had been published: DIEGO CARLSTRÖM (Stockholm), EPHRAIM BANKS (Brooklyn), F. A. HUMMEL (University Park, Pa.), ERIC R. KREIDLER (Cleveland), G. H. MCCLELLAN (Muscle Shoals, Ala.), R. S. MITCHELL (Charlottesville, Va.), E. J. YOUNG (Denver), W. D. ARMSTRONG (Minneapolis) and MARSHALL R. URIST (Los Angeles).

In addition to some of the individuals mentioned above, the editors and/or publishers of several journals have given permission to use previously published diagrams and photographs. The journals involved are: Science (PHILIP H. ABELSON, Editor), the American Mineralogist (WM. T. HOLSER, Editor), Bulletin de la Société Française de Minéralogie et de Cristallographie (C. LÉVY, le Secrétaire général), Archives of Oral Biology (Maxwell International Microforms Corp., Elmsford, New York), and the Geological Society of America Bulletin (Mrs. J. K. FOGELBERG, Managing Editor).

Elimination of errors has been greatly enhanced through the diligent efforts of ADOLF PABST (Berkeley), D. JEROME FISHER (Phoenix, Arizona), and ERNEST G. EHLERS (Columbus); my gratitude to each of them is immeasurable.

I am indebted to a number of co-investigators and to numerous persons who have assisted with laboratory investigations; their names have been mentioned in appropriate places and will not be repeated here. Likewise, it would be quite impossible to indicate the many persons who have contributed to my continuing education in many ways. Included would be mineralogists, chemists, geologists, librarians, physicists, physicians, dentists and engineers—some with academic associations, some in private practice, some with governmental agencies, and some with industrial organizations. Specificly, I wish to mention the late WILLIAM J. McCAUGHEY, my predecessor as chairman of the Department of Mineralogy, and WILFRID R. FOSTER, my successor.

Table of Contents

	page
List of Tables	XIII
Abbreviations	XV
1. Introduction	1
2. Earlier Work on Apatite	5
3. Physical Properties	11
3.1. Density and Dimensions	11
3.2. Optical Properties	12
3.3. Electrical and Other Properties	16
4. Structure	17
5. Crystal Chemistry	22
5.1. Substitutions for Calcium	23
5.2. Substitutions for Phosphorus	26
5.3. Substitutions for Fluorine	29
5.4. Complex Polyionic Substitutions	31
6. Synthetic Apatites: Applications	33
6.1. Phosphors	33
6.2. Synthesis	35
6.3. Phase Relations	37
6.4. Commercial Calcium Phosphate	38
7. Carbonate Apatites	39
7.1. Stoichiometry	39
7.2. Structural Dimensions	42
7.3. Refraction and Absorption	44
7.4. Questionable Interpretations	46
8. Phosphorites	48
8.1. Relations with Sea Water	49
8.2. Geochemistry: Enrichment	50
8.3. Carbonate Content	54

	page
9. Geology: Igneous and Metamorphic Occurrences	56
9.1. Igneous Rocks	56
9.2. Segregations	58
9.3. Veins, Dikes, and Sills	60
9.4. Carbonatites	61
9.5. Metamorphic Rocks	64
9.6. Lunar Rocks and Meteorites	66
10. Biologic Apatites	68
10.1. Teeth and Bones: Composition	69
10.2. Teeth and Bones: Possible Precursors	71
10.3. Teeth and Bones: Mineralization	76
10.4. Teeth and Bones: Trace Elements	77
10.5. Other Biologic Precipitates	79
11. Critique	81
Appendix	87
References	92
Geographical Index and References to Localities	104
Subject Index	107

List of Tables

		page
Table 2.1.	Birefringence values of substances recognized by HAUSEN	5
Table 2.2.	Hypothetical composition compared with that of dahllite	6
Table 2.3.	Densities calculated from the compositions and diffraction measurements of synthetic end members	6
Table 2.4.	Comparisons of morphological frequencies and diffraction intensities	8
Table 2.5.	Forms recognized as occurring on apatite from goniometric measurements	9
Table 3.1.	Influence on the refractive index (omega) by various components	13
Table 3.2.	Calculated refractive indexes (omega)	14
Table 3.3.	Dispersion of apatite	15
Table 4.1.	Atomic parameters for $Ca_{10}(PO_4)_6(OH)_2$	18
Table 4.2.	Comparisons of observed and calculated intensities for synthetic hydroxyapatite	18
Table 5.1.	Comparisons of radii of possible substituents in apatite	23
Table 5.2.	Unit-cell dimensions a for several apatites	24
Table 5.3.	Unit-cell dimensions c for several apatites	24
Table 5.4.	Comparisons of ionic radii for different coordination numbers	26
Table 5.5.	Principal cations of natural apatite varieties	26
Table 6.1.	Synthetic preparations with $A_{10}(ZO_4)_6X_2$ structures indicating Z for various A and X combinations	34
Table 7.1.	Analysis of apatitic wood	40
Table 7.2.	Comparisons of measured and predicted a dimensions for hydrated apatites	43
Table 7.3.	Correlations of composition with unit-cell dimensions and birefringences	44
Table 7.4.	Significant ratios of hydrated carbonate apatites	46
Table 8.1.	Comparisons of abundance of trace elements in phosphorites, crustal rocks and sea water	51
Table 8.2.	Desirability of trace elements in fertilizers	53
Table 9.1.	Comparisons of normative and recovered apatite and other chemical data for different rock types of KIND	57
Table 9.2.	Physical properties compared with certain chemical components for apatites from different rock types	57
Table 9.3.	Analyses of apatites from the Kola Peninsula	59
Table 9.4.	Rare earths and manganese contents of apatites	62
Table 9.5.	Minor constituents of apatites associated with carbonatites	63
Table 9.6.	Minor constituents of sulfate-silicate apatites	65
Table 9.7.	Analyses of apatites from lunar rocks	66
Table 10.1.	Analyses of bovine bone and human dental enamel	69
Table 10.2.	Elemental analyses of human teeth	78

Abbreviations and Symbols

A or Å   angstrom units (10^{-10} m); 1 Å = 1.00202 kX.

a unit-cell dimension (interval of repetition) in the direction of the crystallographic axis a.

b a similar dimension for the b axis.

c a similar dimension for the c axis.

a.w.u.   atomic weight units (herein oxygen is taken as 16.000).

α the crystal angle between b and c axes; the least refractive index or its vibration direction in a biaxial (orthorhombic, monoclinic or triclinic) crystal.

β the angle between a and c; the intermediate refractive index or its vibration direction in a biaxial (orthorhombic, monoclinic or triclinic) crystal which is measured $\perp$ to the optic plane (the plane containing α and γ).

γ the angle between a and b; the greatest refractive index or its vibration direction in a biaxial crystal.

Δ a difference; in connection with optical measurements the birefringence, i.e., $\gamma - \alpha$ or $\varepsilon - \omega$.

d interplanar distance (in Å) of a family of parallel planes in a crystal structure.

EDTA   ethylenedinitrillotetraacetic acid (used as a reagent for chelometric analysis).

ε the refractive index measured with the ray vibrating parallel to c in a hexagonal or tetragonal crystal.

ω the refractive index of the ordinary ray in a hexagonal or tetragonal crystal.

CN   coordination number of an atom or ion, i.e., the number of immediately adjacent atoms with opposite charge.

carHap   carbonate hydroxyapatite, dahllite, containing less than one per cent fluorine.

carFap   carbonate fluorapatite, francolite, containing more than one per cent fluorine.

Cl-*ap*   chlorapatite

Fap   fluorapatite

Hap   hydroxyapatite

Sr-*Fap*   a fluorapatite in which Ca atoms are replaced by Sr atoms.

(hkl) [or ($hk \cdot l$)]   the Miller indices [or Miller-Bravais indices] for a symmetrical family of parallel planes that intercept all three axes of reference; or the interplanar distance (in Å) separating such planes. [The dot represents

	the third index in the hexagonal system where [minus $(h+k)$ is frequently omitted.] A zero indicates there is no interception of a particular axis; $(hk0)$ indicates the plane is parallel with the c axis, for example. In tabulations the parentheses are usually omitted.
m	symbol for a reflection plane of symmetry; in the case of $P6_3/m$ the virgule indicates that it is $\perp$ to the c axis.
MW	molecular weight or, more frequently, the number of a.w.u. contained in the unit cell.
n	nano, indicating 1×10^{-9}. Example: $1 \text{ Å} = 0.1$ nm.
n	refractive index of an isometric crystal or the mean refractive index of a nonisotropic substance.
n.d.	not determined (in connection with chemical analyses).
ppm	parts per million.
R_F	the "ionic refractivity", of a fluorine ion, for example.
RE	a rare-earth atom or ion.
$\rightarrow$	progresses toward or is replaced by.
$\sim$	approaches as a limit, or approximate equality in units or dimensions.

1. Introduction

Apatite is the most abundant of the phosphatic minerals and consequently it is of great importance to industrial chemistry as a raw material. In fact, so many modern technological developments depend upon the use of phosphorus-containing compounds that a fairly complete list would have to include dentifrices, pharmaceuticals, phosphors, rust removers, additives for motor fuels, plasticizers, insecticides and other poisons, and friction matches. The large tonnages, however, are concerned with fertilizers, phosphoric acid and detergents (principally sodium tripolyphosphate). The basic raw material, phosphate rock or phosphorite, is a sedimentary rock of which the essential mineral component is ordinarily a carbonate fluorapatite. Annual production from mining operations in the U.S. increased during 1964—1968 from about 26 to 41 million short tons, and represented between 41 and 47% of world production.

Apatite is a common accessory mineral of most igneous rocks but it is of no commercial consequence in such concentrations. However, pegmatitic and metamorphic rocks may contain concentrations that are of interest, not so much for their phosphorus content as for some of the rarer elements which tend to accumulate in such apatites. The Kola district of the U.S.S.R. contains a rock high in apatite which is related to previous igneous activity, and other rocks of hydrothermal origin may contain appreciable amounts of apatite or francolite (carbonate fluorapatite) as veins or fissure fillings. At Magnet Cove, Arkansas, and in Nelson County, Virginia, apatite is associated with titanium-containing minerals, but it is the latter minerals, rather than apatite, that have attracted commercial interest. An early analysis [238] of apatite from segregated bands in the nelsonite seems to represent ordinary fluorapatite but the summation of oxides is about 1% low and carbon dioxide was not considered.

Association of apatite with iron ores is not uncommon and it may be a source of considerable annoyance with respect to elimination of phosphorus in the finished metal.

Most important from the viewpoint of human welfare are the biologic apatites which comprise the bones and teeth of vertebrates. With the exception of small structures of hard tissue associated with the ear, all normal hard tissues of humans are apatitic, as are many of the pathological calcifications. Stated in its crudest form an important question is: Why do some tissues fail to become mineralized in a normal manner, and why do other tissues that are not mineralized under normal conditions become partially mineralized in pathological situations? The answer to this two-faceted question impinges on a fairly complete knowledge of the metabolic relations which obtain in a "normal" situation in contradistinction to a pathological situation, and thus embraces many aspects of medical science.

Although it is reasonable to believe that the last chapter in medical science is beyond the reach of human endeavor, increments of the organic and inorganic chemistries can be isolated for separate consideration, and factors which interrelate these systems are being sought. At this juncture, however, it must be admitted that a complete knowledge of the crystal chemistry of even the inorganic component of bone has not been obtained.

A chemical compound is not necessarily a mineral, because the term "mineral" is ordinarily restricted to solids, rather than including liquids and gases. Thus a mineral has a particular crystalline structure in addition to other physical and chemical characteristics. The several polymorphic modifications of silicon dioxide are assigned different mineral names which indicate their crystallographic properties, such as refractive indices and electrical characteristics.

While SiO_2 occurs as numerous "mineral" phases, each is more or less characteristic of a particular set of environmental conditions with respect to temperature and pressure. The inversion of low to high quartz, for example, takes place within a few seconds, and although the precise temperature may vary slightly, the range of variation is so small as to permit its use as a reference point for thermal investigations that do not require great accuracy.

Unlike the "chemical compound", a mineral may show a more complex stoichiometry. The plagioclase feldspars are the commonest examples, having as end members, albite ($NaAlSi_3O_8$) and anorthite ($CaAl_2Si_2O_8$). Consequently a mineralogist may be confronted with a common rock-forming substance, such as andesine ($Na_{0.68}Ca_{0.32}Al_{1.32}Si_{2.68}O_8$), which does not have a simple composition until the subscripts are summed in a particular way: $Na + Ca = 1$ and $Al + Si = 4$ for each 8 oxygens. The physical properties, particularly the optical characteristics, are well known for the intermediate compositions so that a fairly reliable prediction of the chemical composition can be made from the optical properties. The essential feature that permits such correlation between physical and chemical properties is the structural similarity.

As an outgrowth of the diffraction methods for investigation of solids, the confines of crystal chemistry were greatly expanded and such topics as *isomorphism* acquired a new significance. The isotypism (same structural type) of berlinite ($AlPO_4$) and quartz ($SiSiO_4$) demonstrated that crystal chemistry had developed new horizons. With respect to apatite, it had previously been found that $(SiO_4) + (SO_4)$ could substitute for $2(PO_4)$ to produce the isotype ellestadite [137].

While the mineralogists were busy attempting to interrelate crystal optics to the newly discovered structures of minerals, physicists and chemists were turning their attention to increasing numbers of "chemical compounds" (i.e., minerals) that were readily available in Nature's laboratory. The fundamental mathematics in terms of description of the 230 space groups already existed prior to any known method of applying this information, but before 1930 the mathematical theory was being applied to the determination of mineral structures, in combination with concepts involving ionic coordination and ionic sizes.

These paragraphs are intended merely as a brief sketch to set in proper perspective the chemical and physical measurements that were soon to be applied to apatite. Personal names, in association with the development of particular theories, have been omitted inasmuch as it is not my purpose to present a history of crystal chemistry, but merely to indicate the background of knowledge that was rapidly becoming applicable to the study of minerals in general, and apatite in particular.

The apatite group received its name (ἀπατάω = I deceive) through WERNER's realization that it was frequently confused with other mineral species, including beryl and tourmaline, prior to the latter part of the eighteenth century. It was early recognized as a calcium phosphate, but its fluorine and/or chlorine content was not known for about forty years.

By 1929, a century later, HAUSEN [90] had brought together considerable information, and had recognized eight different types: 1. fluorapatite, 2. alkali-containing apatite, 3. chlorapatite, 4. carbonate apatite, 5. sulfate apatite, 6. hydroxyapatite, 7. manganapatite and 8. apatite containing rare earths. He states, however, about hydroxyapatite: "Existenz als primäre Verbindung in den eigentlichen Apatiten ist fraglich", and thereby questions the existence of such an end member.

The work of HAUSEN [90] antedates a knowledge of the structure, of course, and considers the chemical compositions within the framework of the concept of "double salts", that is, "molecules" of tricalcium phosphate in combination with "molecules" of calcium fluorite to give "molecules" of fluorapatite. The more amazing fact is that such concepts persist to the present day with complete indifference toward the more realistic methods for formulating isomorphic substitution[1].

Although these introductory remarks are not intended to supply basic information on modern crystallochemical concepts, it must be appreciated that some of the more fundamental principles were first discovered through investigation of apatite. For example, the discovery of ellestadite [137] was considered of sufficient significance to receive commentary by the famous Russian mineralogist, VERNADSKY [227]. Although aluminum had long been recognized as an atom capable of occupying sites of silicon, particularly in feldspars, its ability to substitute for both Ca and P in an orthophosphate, and in significant amounts, represents a recent extension of the possibilities of isomorphic substitution [62].

Indeed, with the exceptions of the clay minerals and the feldspars probably no other single mineral group has proved as difficult to resolve as the apatites. Part of the difficulty has stemmed from encroachment into this area by persons

---

[1] An example of interest is the garnet isotype sometimes called tricalcium aluminate hexahydrate. This name might be taken to imply that 6 water molecules are present in the structure ($3CaO \cdot Al_2O_3 \cdot 6H_2O$) whereas the structure is characterized by 12 hydroxyl ions occurring as (H_4O_4) groups, analogous to (SiO_4) groups of grossular [66]. Indeed, the writing of mineral compositions in the form of oxides leaves something to be desired; apatite becomes $5CaO \cdot 1.5P_2O_5 \cdot F$ from which must be deducted half an atom of oxygen in order to obtain electrical balance. One could write for apatite $9CaO \cdot CaF_2 \cdot 3P_2O_5$, but the expression of a single crystalline phase as a combination of two or more nonisotypic phases has nothing to recommend it and has contributed to serious confusion.

Under special circumstances, such as the case of dolomite [$CaMg(CO_3)_2$], the expression "double salt" might have some significance inasmuch as dolomite is not an isomorphic variant between the calcite and magnesite isotypes, but has a structure characterized by ordering of the Ca and Mg atoms.

quite inadequately prepared in crystallography and inorganic chemistry, with the effect that it was even proposed that the citrate content of bone might enter the "lattice of bone mineral"!

As in the case of some clay minerals, the individual crystallites of bone do not lend themselves to single-crystal diffraction techniques. Consequently, it becomes necessary to apply indirect methods and analogies to closely related substances: francolite and dahllite for bone, micas for clay minerals. In neither situation was there a dearth of observational data — both good and bad — but deductive methods were essential to the organization of these bodies of empiricism into science.

Both topics (clays and biologic apatites) suffered reverses at the hands of "absorptionists" who refused to recognize that fundamental crystallochemical principles were involved, and who attempted to apply colloid science in inappropriate ways.

In the chapters that follow no attempt will be made to cover all of the descriptive accounts of the occurrences of apatite, haphazard attempts at synthesis that are not adequately supported by analytical data, x-ray diffraction data devoid of evidence of calibration or use of internal standards, and numerous experiments — particularly some involving infrared absorption — that have contributed nothing.

It is one thing to determine the crystal structure of a solid, but it is another thing to know the chemical composition of the same solid. To be sure, this statement may seem somewhat naive, but its naivete is not as great as many reports on apatite that have appeared in the literature — reports that made pretense of superior sophistication because of the use of intricate and expensive apparatus and/or computation programs, but which failed, nevertheless, to take into account significant data obtained by other methods.

It must be recognized that one chemical analysis made by reliable methods is far more useful than a hundred analyses of dubious quality, and that precision obtained by a faulty method in no way contributes to accuracy. Inasmuch as crystallochemical deductions rely heavily on analytical determinations, it becomes essential to consider the probability of error in view of the analyst's skill, knowledge and experience. For example, a reliable method for fluorine was not developed prior to 1933 [241, 8] so that earlier values must be regarded with suspicion. Even today, there is uncertainty that all of the water content of an apatite can be obtained by a direct method, and such reports as "ignition loss" are quite unsatisfactory for most purposes.

In general, the present work will not become concerned with analytical methods for apatites, except when some crucial determination is involved. Analysis of phosphates is more difficult than that of ordinary silicate rocks, and the National Bureau of Standards and the U.S. Department of Agriculture have devoted considerable attention to this topic. CRUFT et al. [41] have outlined some of the procedures particularly applicable to apatite and spectrochemical methods have been discussed by WARING & CONKLIN [237].

2. Earlier Work on Apatite

As mentioned, HAUSEN [90] attempted to summarize some of the data and conclusions in 1929, but his work suffered from two severe disadvantages: the structure had not been determined and many of the analyses at his disposal were not reliable with respect to their fluorine contents.

It was recognized that the refractive index (ω) was influenced by the composition, increasing by about 0.005 for each weight per cent of chlorine of the analysis. For these purposes the indexes (ω for Na light) of fluorapatite and chlorapatite were taken as 1.6325 and 1.6667, respectively, from synthetic preparations. Likewise, the effects of MnO and FeO were evaluated as increasing the refractive index by 0.0030 and 0.0040 per weight per cent of these oxides.

The birefringence was also recognized as being influenced by the composition but these relations were less definitive, and were derived from some misconceptions concerning structural compositions. Nevertheless, HAUSEN's data on several materials are interesting (Table 2.1). Numerous other values for birefringence were

Table 2.1. *Birefringence Values of Substances Recognized by* HAUSEN
(The Substances are Quoted Directly)

Substance	$\omega - \varepsilon$	Composition (% wt.)
Synt. CaCO$_3$: Apatit	0.0090	CO$_2$ 4.2
Synt. NaCl: Apatit	0.0058	Cl 4.85; Na$_2$O 2.11
Synt. CaCl$_2$: Apatit	0.0053	Cl 6.8
Synt. CaF$_2$: Apatit	0.0030	F 3.8
Åbo, Finland	0.0016	F 1.95; Cl 0.47; Na$_2$O 2.05

listed for apatites from various localities, but even from the data just supplied an anomaly arises in terms of the variety from Åbo. It contains approximately the same amount of Na$_2$O as "Synt. CaCl: Apatit" but no amount of F or Cl that could account for its very low birefringence. The birefringence of chlorapatite is now known to be less, rather than greater, than that of fluorapatite.

Although it was recognized that carbonate groups were capable of significantly increasing the birefringence, it was supposed that the composition of the synthetic material prepared by EITEL had the composition Ca$_9$(PO$_4$)$_6$ · CaCO$_3$ and therefore contained the theoretical amount of carbon dioxide. On the other hand, a theoretical composition was given for an acid carbonate phosphate which was compared with dahllite (Table 2.2). In the absence of an absolute measurement of the unit-cell volume, it was not recognized that such an "acid carbonate phosphate" would necessarily involve a considerably greater theoretical density. That is, the 2 F atoms would weigh 38 units whereas the $2(HCO_3)$ would weigh 122.

Table 2.2. *Hypothetical Composition Compared with That of Dahllite*

$3Ca_3P_2O_8 \cdot Ca(HCO_3)_2$		Dahllite, Kangerdluarsuk	
CaO	51.30	CaO	54.10
P_2O_5	39.00	P_2O_5	32.40
CO_2	8.05	CO_2	8.26
H_2O	1.65	Na_2O	0.77
	100.00	Al_2O_3	3.15
		H_2O	1.32
			100.00

Table 2.3. *Densities Calculated from the Compositions and Diffraction Measurements of Synthetic End Members*

Composition	Volume (Å³)	Density	
		Calculated	Observed
$Ca_{10}(PO_4)_6(OH)_2$	528.5[a]	3.1559	—
$Ca_{10}(PO_4)_6F_2$	523.4[b]	3.1992	3.180
$Ca_{10}(PO_4)_6Cl_2$	545.5[c]	3.1695	3.170

[a] McConnell & Foreman [160].   [b] Bhatnagar [16].   [c] Bhatnagar [17].

With the advent of x-ray measurements it was discovered for francolite, that not only was the unit-cell volume smaller but the observed density was also less, thus completely precluding the possibility of substitution of even one CO_3 group for 2 F to give a carbonate apatite. In addition, francolite from Staffel, Germany, contained more than the required amount of fluorine [80].

Specimens tabulated by Hausen indicate a range of specific gravities from 3.410 for a manganese-containing variety from India to 3.060 for a "voelckerite" from Santa Clara County, California.

It is interesting to compare densities calculated from the volumes of the unit cells and the "molecular" weights with the earlier measurements on synthetic substances reported by Hausen (Table 2.3). The "voelckerite", were it to have such a composition as $Ca_{10}(PO_4)_6O$, could not have a volume in excess of hydroxyapatite (528.5) so the minimum calculated density would be 3.0994; thus it must be assumed either that the measurement on the Santa Clara County specimen involves an unusually high error or that the chemical composition is significantly different from that given by Rogers [197]: CaO 54.46, Al_2O_3 1.35, FeO 0.24, P_2O_5 41.47, H_2O 0.22, CO_2 1.03, Insol. 0.53/99.30%.

Numerous measurements with the two-circle goniometer were available to Hausen [90], as also was Goldschmidt's "Atlas" [74], from which as also was some more or less typical illustrations are reproduced here (Figs. 1 to 7). The habit is usually prismatic, on occasion almost needlelike, but the tabular habit is known also. The axial ratios compiled for natural crystals ranged from 0.7357 to 0.7246, whereas two much lower values are given for synthetic chlorapatite, namely 0.7028 and 0.6983.

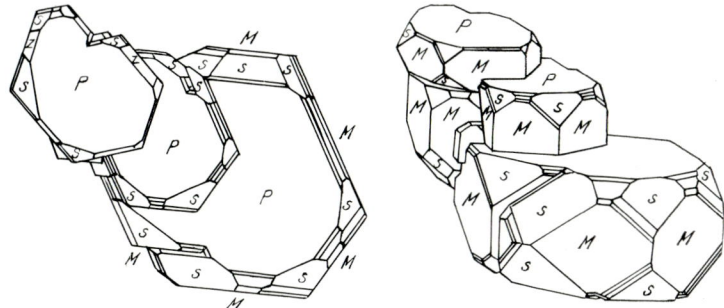

Fig. 1. Group of tabular crystals of apatite (plan and clinographic views) showing P (00.1), M (10.0) and s (11.1). (Goldschmidt [74])

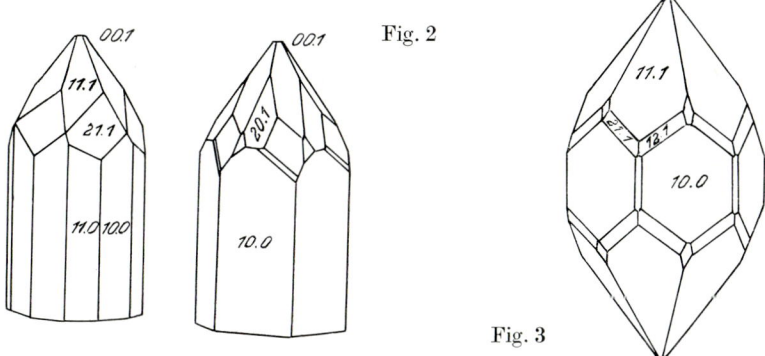

Fig. 2. Two prismatic crystals showing prominent pyramidal forms (11.1) and (21.1) (Fig. at left) and the additional form (20.1) (Fig. at right). Both prisms (11.0) and (10.0) appear on the left, but merely (10.0) on the right. The pinacoidal faces (00.1) are rudimentary on both examples. Examples are from Pfitsch in Tirol. (Goldschmidt [74])

Fig. 3. Another example from Pfitsch (Tirol) showing terminal faces above and below, as well as absence of (00.1). The prominent prismatic form is (10.0), whereas the prominent dipyramid is (11.1), and (21.1) and (12.1) are shown. (Goldschmidt [74])

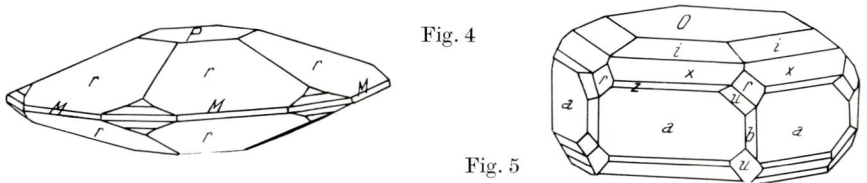

Fig. 4. Example from Cavorgia (Tavetsch) showing the dipyramid r (10.2), as well as P (00.1) and M (10.0). (Goldschmidt [74].) Crystals with similar habit occur near Hebron, Maine [185a]

Fig. 5. Tabular crystal from St. Michaels Mt. (Cornwall) showing: O (00.1), a (10.0), z (20.1), x (10.1), i (10.2), b (11.0), r (11.1) and u (21.1). (Goldschmidt [74])

Recent measurements by x-ray diffraction indicate for chlorapatite [17] $c = 0.702$, for hydroxyapatite [160] $c = 0.731$, and for fluorapatite [16] $c = 0.735$; thus the entire range of morphologic measurements are encompassed with the exception of the extreme high and the lower of two ratios for chlorapatite. An attempt to correlate birefringence with axial ratios produced nothing of value

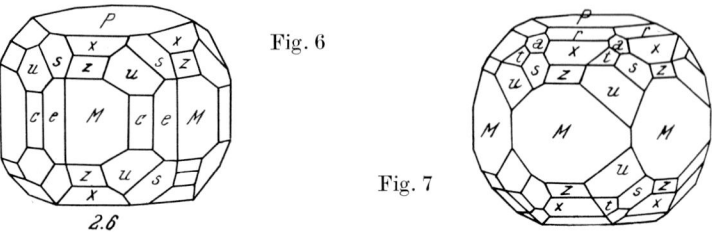

Fig. 6. Crystal from Schlaggenwald showing: P (00.1), M (10.0), c (21.0), e (11.0), u (21.1), s (11.1), z (20.1) and x (10.1), in approximately decreasing order of prominence. (GOLDSCHMIDT [74])

Fig. 7. Crystal similar to Fig. 6., but showing additional pyramidal forms: a (11.2), r (10.2) and t, which seems to occur in the same zone; thus presumably t is (21.4). (GOLDSCHMIDT [74])

because of the number of variables affecting both properties, as well as the experimental difficulties involved in obtaining accurate values for birefringence. The birefringence for synthetic chlorapatite, taken as 0.0053 by HAUSEN, is quite unrealistic, for example.

Comparisons of the frequencies of occurrences of faces or forms, as observed on external morphology, with intensities observed by x-ray diffraction are interesting. Quantitatively, but on the basis of merely 16 crystals from 6 localities, the frequencies of occurrence of faces (for any particular form) are compared (Table 2.4)

Table 2.4. *Comparisons of Morphological Frequencies and Diffraction Intensities*

Order of morphologic frequency $hk.l$	Powder diffraction intensities[1]				Multiplicity factors
	1st order	2nd order	3rd order	4th order	
00.1	abs.	46	abs.	(x)	2
21.1	100	—	—	—	24
10.0	w.	w.	39	abs.	6
10.2	18	—	—	—	12
10.1	w.	23	—	—	12
11.1	w.	15	—	—	12
20.1	abs.	(x)	—	—	12
11.0	(x)	abs.	abs.	—	6
31.1	w.	—	—	—	24
11.2	34	—	—	—	12
30.1	w.	—	—	—	12

[1] Relative to 100 for the greatest intensity; *w.* indicates an intensity less than 10; abs. indicates absence; a dash represents a larger angle and consequent uncertainty; (x) not reported for apatite from Durango [252] but frequently occurs for crystals from other localities.

2. Earlier Work on Apatite

with the intensities of powder diffraction maxima for a crystal from Cerro de Mercado, Durango, Mexico [252]. All crystals measured with the two-circle goniometer showed the basal pinacoidal form. The next commonest form was (21.1) — in some cases both (21.1) and (12.1) — so it is not surprising that this should be the most intense diffraction line.

Considering the Lorentz-polarization and multiplicity factors in combination with characteristic absences produced by symmetry elements (for example, the 6_3 axis) there is only one obvious anomaly, namely absence of the first-, second- and third-order reflections of (11.0). However, this is not as disturbing as it might seem, because the reflection (11.0) has been reported for synthetic hydroxyapatite [160] and for fossil bone [24], albeit with very low intensity.

Considering the data from the opposite viewpoint causes difficulty in accounting for the ten most intense powder diffraction maxima on the basis of the geometrical frequencies, particularly (21.3) and (31.0), with respective intensities of 28 and 15. Comparison of powder diffraction patterns from other localities does not alleviate the situation, so it must be concluded that the correlation between morphologic frequency and diffraction intensity is only moderate, possibly because of insufficient observations, although the multiplicity factors are large: 24 for (21.3) and 12 for (31.0), including (13.0).

In addition to the basal pinacoid (00.1) at least 29 forms are represented on apatite, according to GOLDSCHMIDT [74] (Table 2.5).

Table 2.5. *Forms Recognized as Occurring on Apatite from Goniometric Measurements*

Prismatic	Pyramidal				
10.0	10.1	11.1	20.1	30.1	40.1
11.0	10.2	11.2	21.1	30.2	50.12
21.0	10.3	11.4	21.2	30.4	70.3
41.0	10.6	11.6	22.1	30.5	[1]
		11.12	31.1		

[1] Less common forms include: (41.1), (43.1), (31.2) and (73.3).

Thus far, the survey of earlier work on apatite has been confined to that presented by HAUSEN [90], but the summary by FORD [65] in 1932 surely deserves attention. Here a partial recognition of hydroxyapatite [$Ca_4(CaOH)(PO_4)_3$] is implied, as well as voelckerite [$Ca_4(CaO)PO_4)_3$[sic]]. The physical properties for fluorapatite given by FORD follow:

Cleavage: c (0001) imperfect; m (10$\bar{1}$0) more so.
Brittle, hardness 5 (sometimes 4.5 when massive).
Specific gravity (G.) = 3.17—3.23. Luster vitreous, inclining to subresinous. Color usually sea-green; bluish green; often violet-blue; sometimes white; occasionally yellow, gray, red, flesh-red and brown. Optically negative. Birefringence low. $\omega = 1.632—1.648$, $\varepsilon = 1.630—1643$.

This maximum for ω (1.648) presumably is based on measurements made before 1905 on material from Pisek (Bohemia) which was reported to have a surprisingly low density (3.094) for any apatite variety. The analysis presumably is that of an ordinary fluorapatite with traces of Cl and MnO; surely the moderately high refractive index and the very low density are completely inconsistent with the analysis inasmuch as the values for synthetic fluorapatite are approximately 1.633 and 3.20.

Besides the several criteria used for recognition of apatite, FORD discusses numerous types of geological occurrences and localities, including the observation of apatite "as the petrifying material of wood". He includes under apatite several varietal names: eupyrchroite, nauruite, phosphorite, staffelite, francolite, dahllite, kurskite, podolite, moroxite, fermorite, wilkeite and svabite. Dehrnite and lewistonite are mentioned elsewhere, but about the former he states: "May be a member of the apatite group."

With adequate justification, several of these varietal names have been abandoned, including quercyite (= dahllite), staffelite (= francolite), eupyrchroite, nauruite, kurskite, podolite and moroxite. Collophane (collophanite of FORD) is used as a petrographic term to indicate a cryptocrystalline carbonate apatite for which the fluorine content is not known [139]. Voelckerite is a synonym for oxyapatite; other synonyms are given [185a]. The distinction between francolite and dahllite depends on the fluorine content (greater or less than 1%, respectively) [138]. Phosphorite should be applied as the name for a rock composed essentially of phosphates of calcium that might include other minerals (whitlockite and brushite, for example) [143, 144].

FORD lists 7 forms likely to occur on apatite, namely: c (00.1), x (10.1), y (2.10), r (10.2), s (11.1), m (10.0) and μ (21.1). Eight interfacial angles are given.

Many of the earlier observations were summarized by FRONDEL [70] in 1951, and need not receive further consideration here, except with respect to twinning. Pyramidal faces (11.1) and (10.3) are reported as rare twinning planes, as also are (11.3) and (10.0). It should be pointed out, however, that both $(hk \cdot l)$ and $kh \cdot l)$ forms are known to occur on apatite and their simultaneous occurrence, particularly of (21.1) and (12.1), could be interpreted as twinning on (10.0), i.e., a 2-fold axis perpendicular to the first-order hexagonal prism.

3. Physical Properties

It may be wondered why a separate chapter should be devoted to physical properties when this entire work is concerned, either directly or indirectly, with the physics and chemistry of apatite, but it does not seem inappropriate to examine some of the data from a general, theoretical viewpoint.

3.1. Density and Dimensions

Density (or specific gravity) is a difficult property to measure for substances that contains gas, liquid or solid inclusions. Apatite frequently contains microscopic inclusions, sometimes to such an extent that it loses its transparency through cloudiness produced by these extraneous substances. Numerous measurements of the density of apatite have appeared in the literature, and one that appears to have involved considerable care [252] has been obtained for the transparent crystals from Cerro de Mercado, near Durango, Mexico:

(Grams per cm³ at 22° C) Measured 3.216 ± 0.002; calculated 3.219.

This, of course, does not correspond with $Ca_{10}(PO_4)_6F_2$ because analysis of the crystals from Durango departs somewhat from this composition.

Calculation of the density from the measured unit-cell dimensions for synthetic fluorapatite depends primarily upon the accuracy with which Avogadro's number is known, provided the composition is assumed to be $Ca_{10}(PO_4)_6F_2$. That is, the accuracy of the unit-cell volume can be excellent and the accuracy of the atomic weights do not impose serious problems. Assuming that $a = 9.3680$ and $c = 6.8868$ A [16] and that Avogadro's number is 6.024×10^{23} one obtains 3.1988 g/cm³ for fluorapatite (actually 3.199 inasmuch as the error is surely in the fourth significant figure). For measurements, the error frequently may be in the third significant figure.

Whereas the calculation depends on an accurate knowledge of the composition, as well as the unit-cell volume, there are indirect methods for estimating the volume solely from the composition within an isotypic series. Using such a method [155], the predicted volume for vanadinite is 673.6 A³, as compared with 678.7 A³ obtained by TROTTER & BARNES [218].

In the case of natural apatites that contain very little chlorine there is comparatively little variation of the dimension c which approximates 6.88 A to within about 0.015. Thus, a prediction of the a dimension may be made, if this relation is assumed to be strictly additive, through use of the following equation [158]:

$$a_p = a_k + x(A) + y(B) + \ldots \tag{1}$$

where

a_p	is the predicted dimension (in A),
a_k	is a constant inherent to the structure,
A, B	are numbers of atoms of different species, and
x, y	are coefficients related to the sizes of the particular atoms.

Necessarily the constant must lie between *Fap* and *Hap* and has been estimated as 9.404 A, to yield predictions of 9.374 and 9.419 as compared with 9.368 and 9.416 for respective values measured on synthetic materials. The coefficients assigned for the *compositional* model were: hydrogen 0.0075, fluorine —0.015, chlorine 0.115, and carbon —0.70. Applications of such calculations are discussed further under carbonate apatites.

Such predictions involving substitution for Ca by Sr, Pb, etc. are comparatively simple, and dimensional differences for such purposes are given later (Tables 5.2 and 5.3). COLLIN [38] has shown that this relation for Ca →Sr is essentially linear; for a and c changes are about 0.04 A per atom substituted.

Calculations of these sorts are necessarily limited in their applicability because of changes in site or differences in coordination number. Besides taking into account the substitution of V for P in vanadinite, the first relation fails to recognize that Cl does not occupy the structural position of F and has, as a consequence, a different coordination number. Likewise in Equation 1, it should be apparent that the coefficient for H is actually the average effect of an additional proton without consideration of whether it is combined to produce an OH ion or an OH_3 ion. One recalls that some francolites contain more F than $Ca_{10}(PO_4)_6F_2$, so the compositional model, $Ca_{10}(PO_4)_6X_2$ is, in a sense, that of oxyapatite which is not electrically balanced, and the coefficients for H, F and Cl are functions of the differences between their radii and the radius of oxygen, except that H is always assumed to be bonded to oxygen and therefore to cause an increase in size.

It becomes evident from the above considerations that an accurate knowledge of the chemical composition becomes essential, and while polymorphism completely precludes obtaining a detailed knowledge of the structure from composition, a determination of structure lacks significance if the composition is not *accurately* known. It follows, therefore, that a "refinement" of the structure of *Hap* or *Fap* is virtually meaningless if the intensity maxima were obtained from a substance which did not accurately correspond with $Ca_{10}(PO_4)_6(OH)_2$ or $Ca_{10}(PO_4)_6F_2$.

3.2. Optical Properties

One of the properties most frequently measured is the refractive index or indices. With the use of several precautions, such measurements can attain an accuracy of ± 0.001, provided that the immersion oil is calibrated at the same time and same temperature on a refractometer.

In the discussion of early work on apatite are mentioned some of the conclusions of HAUSEN [90] that attempted to correlate optical properties with compositional differences. It is not surprising that little success was attained in these earlier investigations, however, because the refractive index is particularly sensitive to compositional changes of certain types, and HAUSEN's compositions were not accurately known.

The relation of GLADSTONE & DALE — devised for solutions, rather than solids — has been applied indirectly to apatite by YOUNG et al. [251, 252] through use of 1.6325 as the omega index for *Fap* and incremental increases which result from substitutions of one weight per cent of each component (Table 3.1). Recent results obtained by TABORSZKY [213a] are compared with those of YOUNG in Table 3.1. The relation of GLADSTONE & DALE, however, is probably viewed with less favor than the relation of LORENTZ & LORENZ for calculations of this sort.

Table 3.1. *Influence on the Refractive Index (Omega) by Various Components (Per Weight Per Cent)*

Component	YOUNG	TABORSZKY
H_2O	0.0091	0.008
Cl	0.0054	0.005
FeO	0.0040	0.004
RE oxides	0.0018	0.003
MnO	0.0020[a]	0.003
SO_3	0.0003	−0.014
SrO	0.0001	—
Na_2O	—	−0.004
F	—	−0.004
CO_2	—	−0.007

[a] The value given [251] is 0.0019_8.

McCONNELL & GRUNER [163] concluded that decreases in the mean refractive index might occur with entry of Na, Mg and C ions into the structure but that increases would occur for K, Sr, Mn, RE, Cl, and OH. Their principal objective was demonstration that carbonate groups were the only mechanism for significantly increasing the birefringence and simultaneously decreasing the mean index, which they estimated to be between 1.631 and 1.634.

The application of the theory of LORENTZ & LORENZ involves difficulty because the mean refractive index (n) is a function of the density as well as the summation of the ionic refractivities. However, the less-accurate measurement of density can be eliminated by substitution of the product of the volume and Avogadro's number to give:

$$\frac{n^2-1}{n^2+2} = \frac{1.660}{\text{vol.}} (X+Y+Z+\ldots) \qquad (2)$$

where X, Y, Z, etc. are products of the ionic refractivities and the numbers of ions for each elemental species. For *Fap* the summation $(X+Y+Z+\ldots)$ becomes 112.5 and for *Hap* and *Cl-ap* 116.3 and 121.5. The differences, $R_{OH} - R_F = 1.9$, $R_{Cl} - R_{OH} = 2.6$ and $R_{Cl} - R_F = 4.5$, will permit some averaging in an attempt to estimate the ionic refractivities of the ions of the three common apatites:

Ca	P	O (of PO_4)	F	OH	Cl
2.10	0.25	3.5	3.0	4.9	7.5

Although R_{OH} is the same as that found for $Ca(OH)_2$, the value for F is considerably larger than that for simple fluorides (2.20), and that for Cl is considerably smaller than that for simple chlorides (8.45). The contribution of the 24 oxygens was assumed to remain the same for the three structures, and the mean R_{Ca} was not corrected for coordination number; in garnets [151] the value for R_{Ca} was estimated as 2.08, whereas it has been estimated to be 1.99 for carbonates.

Using the LORENTZ-LORENZ relation, BIGGAR [19] encountered similar uncertainties and obtained results somewhat different from those of YOUNG and his coworkers, mentioned previously. However, BIGGAR did not attempt to assign refractivities to the several different ions, as done here, but was concerned with plots of the weight percentages of several oxides and halogens versus the omega indexes, as had been done previously by KIND [110] and others.

The unit-cell volumes used here are slightly different from those used by BIGGAR; they are as follows:

	Fap	*Hap*	Cl-*ap*
Ca	523.1	528.7	544.5
Sr	594.4	600.6	606.6

A natural apatite containing 46.06% of SrO is reported [248] to have $\omega = 1.651$ and $\varepsilon = 1.637$, but it also contains other oxides, such as those of Ca, Ba, Th and RE, and its birefringence is unaccountably high (0.014). Although its volume is given (581.0 A^3), lack of information concerning appropriate ionic refractivities of several minor components precludes a calculation of its refractive index through the use of Equation 2.

In summary, it is interesting to compare the omega indexes when calculated by different methods and compare them with available data on synthetic products (Table 3.2). For the purpose of these calculations R_{Sr} was estimated to be 3.41, and the volumes were those given above. A measurement of ω for Sr-*Hap* does not seem to have been recorded despite several reported syntheses. Thus, while R_{Sr} exceeds R_{Ca} by a factor of about 1.6, it is deduced that the size of the Sr ion is sufficiently greater to reduce the refractive index rather than raise it. BIGGAR's

Table 3.2. *Calculated Refractive Indexes (Omega)*

	This work	YOUNG	BIGGAR	Observed	Reference
Fap	1.632	1.6325	1.632	—	—
Hap	1.651	1.6488	1.646	1.651	[160]
Cl-*ap*	1.666	1.6693	1.700	1.668	[21]
Sr-*Fap*	1.619	1.6395	1.618	1.621	[243]
Sr-*Hap*	1.634	1.6406	1.628	—	—
Sr-Cl-*ap*	1.658	1.6699	1.675	1.658	[33]

values are consistent with this conclusion insofar as his Sr-containing apatites have indexes lower than those of the corresponding Ca-containing isotypes. For Young's calculated indexes, only that of Sr-Hap is lower than Hap, although the index for Sr-Cl-ap is almost the same as for Cl-ap. Merely one conclusion emerges: The refractive index of apatite, while clearly coupled with its composition, involves complex electromagnetic vectors the nature of which will not be easily resolved through the data now available. A similar situation appears to arise with respect to vibrational frequencies in the infrared range [111].

Dispersion of apatite is sufficiently great to require specification of the wavelength and to produce doubt concerning the accuracy of measurements obtained using sunlight or incandescent tungsten without filtering. Baumhauer [14] has measured dispersion (Table 3.3). One notes, however, that his measurement (ω) at 588 nm [Na flame approximates 589 nanometers (1 nm $= 1 \times 10^{-9}$ m)] is somewhat above that of ordinary Fap, and $\Delta = 1.6388 - 1.6357$ is slightly less than that ordinarily attributed to Fap.

Table 3.3. *Dispersion of Apatite*

Wavelength (nm)	ω	ε
447	1.6497	1.6462
502	1.6445	1.6413
588	1.6388	1.6357
688	1.6356	1.6326

Apatites that are colored may show pleochroism, with the absorption greater for epsilon than for omega. Besides zonal distribution of coloration, which is known for pegmatitic apatites containing manganese, the carbonate apatites frequently are biaxial, and do not show parallel extinction in basal sections. The latter statement is particularly applicable to cross sections that exhibit six sectors surrounding a central core [45], but polysynthetic twinning with the hexagonal prism as the composition plane occurs also [163].

Coloration of apatite is fairly common, greenish yellow probably being the most typical. However, pale pink, lavender and orange are not rare. Blue apatite from near Keystone, South Dakota, has been found [106] to owe its color to the presence of the MnO_4^{-3} ion. The natural color was bleached by heating for a few minutes at 600° C, but this was not true for blue synthetic apatites. Manganese-containing apatites are usually colored and the intensity, as well as the tint, is related to the amount of Mn, as well as its state of oxidation [244].

Hoffman [96] attributed green coloration to the simultaneous presence of ferric and ferrous ions, but there is doubt that the color of apatite from Cerro de Mercado, Durango, Mexico, can be attributed to this cause inasmuch as ferrous ions were present to an extent less than 100 ppm, whereas rare earth oxides were 1.43% by weight [252]. The fluorescent properties, and probably coloration also, are known to be related to rare earth contents [84].

Grisafe & Hummel [79] have discussed in considerable detail the effects of Co, Ni, La, Pr and Nd, and deduced that the spectral absorption bands for Ni and

Co were similar to those produced by these ions in octahedral coordination. Their investigations involved silicates (in addition to phosphates) with the apatite structure that included Pb and Sr in addition to rare earths.

3.3. Electrical and Other Properties

The dielectric constant (i.e., the specific capacity of a material referred to the specific capacity of a vacuum) is given [35, p. 566] as ranging from 7.40 to 10.47 for radiofrequencies. For a crystal from Asio, Japan the measurement parallel with c was 10.0 and perpendicular to c was 7.60. For optical frequencies the respective values were 2.71 and 2.69. Additional data are given by WAPPLER [236]. Theoretical discussions of the low dielectric loss of hydroxyapatite are not edifying insofar as they are based on an inadequate "refinement" of the location of the OH ions within the structure [253].

Piezoelectric properties are not consistent with the space group $P6_3/m$, of course, so the attribution of such properties to bone implies that $carHap$ has a less symmetrical structure; if piezoelectric properties are to obtain for the aggregation, the individual crystallites must have piezoelectric characteristics. This topic will receive further consideration under biologic precipitates.

Magnetic susceptibility, measured for crystals from Cerro de Mercado, Durango, Mexico, by SENFTLE (reported [252]) indicated very low values at ordinary temperatures. At 27° C this apatite was found to be diamagnetic, presumably because of its very low iron and manganese contents.

The melting points for synthetic apatites have been determined by use of a hot-stage microscope [15] with these results (two determinations each):

Hap	1614°, 1614° C
Fap	1615°, 1622°
Cl-*ap*	1612°, 1612°
Sr-*Hap*	1670°, 1670°

Two indistinct cleavages, one parallel and one perpendicular to the basal pinacoid, may influence to some extent the relative intensities of diffraction maxima because of preferential orientation of grains during preparation of x-ray patterns by the powder method. The more characteristic fractured surface of apatite is conchoidal without good cleavage. By definition, fluorapatite is five in Mohs' arbitrary scale for scratch hardness. However, some carbonate apatites, depending upon their water contents, have been reported as having hardness as low as three. HODGE [94] applied hardness tests to teeth and recognized some of the problems connected with determination of hardness for substances composed of aggregations of tiny crystals.

The mechanical properties of apatite are of little interest except in connection with study of the skeletal structure of vertebrates, and such biologic apatites are significantly different with respect to their crystal chemistry. Thus, an attempt to infer the characteristics of bone from measurement of the elastic properties of "gem-quality" crystals from Durango [250] seems somewhat farfetched. Values for compressibility and elastic constants are given in the compilation by CLARK [35].

4. Structure

The space-group symmetry of apatite was indicated as $P6_3/m$ or C_{6h}^2 by HENTSCHEL [93] in 1923, by the use of von Laue and rotation photographs, and the essential features of the structure were determined independently by NÁRAY-SZABÓ [180] and MEHMEL [174], both of whom published their results in 1930.

Numerous attempts to refine the parameters given in these early investigations have not brought about spectacular improvement inasmuch as these so-called refinements have been performed on unanalyzed crystals. A set of approximate parameters is given in Table 4.1; these are generally applicable to both fluorapatite and hydroxyapatite. Fig. 8 is the corresponding diagram.

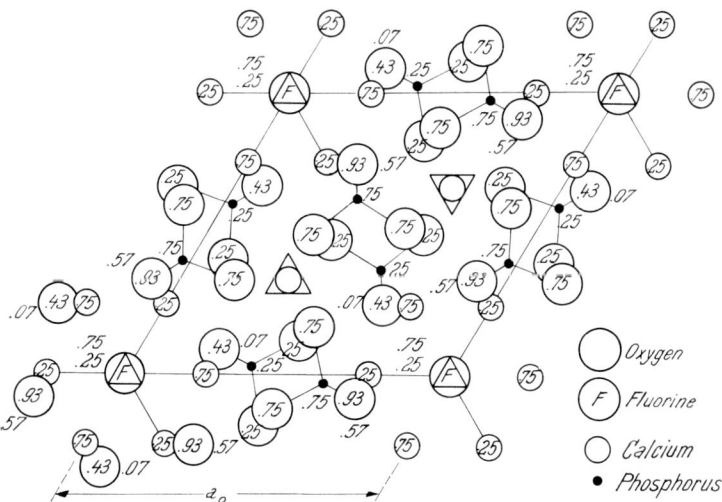

Fig. 8. Projection of the apatite structure on (0001). Elevations are represented as fractions of c; m planes are located at 0.25 and 0.75 c; △ represents 3-fold axes on which Ca atoms are located; the origin is the 6_3 axis (lower right). (McCONNELL [138])

Intensity measurements are given in Table 4.2, where they are compared with calculated intensities which assume symmetric distribution of the electron densities for OH^- across the m planes. The Ca/P ratio of this synthetic preparation closely approached 1.67 and very little carbonate was present. The refractive indices were $\varepsilon = 1.647$ and $\omega = 1.651$, and $a = 9.416$ and $c = 6.883$, both $\pm 0.002\,\text{Å}$ [161].

In connection with the OH group, its lack of symmetry raises a question concerning the electron density astride the m planes (at $1/4$ and $3/4\,c$) which YOUNG &

4. Structure

Table 4.1. *Atomic Parameters for* $Ca_{10}(PO_4)_6(OH)_2$

Atom	x	y	z
Ca_I	0.33	0.67	0
Ca_{II}	0.25	0	0.25
P	0.40	0.37	0.25
O_I	0.33	0.50	0.25
O_{II}	0.60	0.48	0.25
O_{III}	0.35	0.26	0.07
OH	0	0	0.25

Table 4.2. *Comparisons of Observed (I_o) and Calculated (I_c) Intensities for Synthetic Hydroxyapatite*[1]

d_o	d_c	hk·l	I_o	I_c	d_o	d_c	hk·l	I_o	I_c
8.147	8.158	10.0	10	15	2.229	2.228	22.1	3	2
5.267	5.259	10.1	5	8	—	2.208	10.3	—	<1
4.711	4.710	11.0	2	<1	2.148	2.149	31.1	5	7
							13.1		
4.079	4.079	20.0	6	12	—	2.133	30.2	—	<1
3.892	3.887	11.1	6	<1	2.060	2.062	11.3	4	7
3.506	3.509	20.1	5	7	2.039	2.039	40.0	2	2
3.441	3.440	00.2	35	34	1.998	1.999	20.3	5	2
3.175	3.170	10.2	10	12	—	1.955	40.1	—	1
3.083	3.083	21.0	19	14	1.943	1.943	22.2	34	32
		12.0							
2.812	2.814	21.1	103	98	1.891	1.890	31.2	13	16
		12.1					13.2		
2.778	2.778	11.2	65	43	1.871	1.872	32.0	5	6
							23.0		
2.718	2.719	30.0	74	61	1.845	1.840	21.3	44	32
2.628	2.630	20.2	26	24			12.3		
2.525	2.529	30.1	5	4	1.805	1.806	32.1	19	20
							23.1		
—	2.355	22.0	—	<1	1.779	1.780	41.0	18	13
							14.0		
2.297	2.296	21.2	4	8	1.755	{1.754	40.2}	16	{15
		12.2				1.753	30.3}		2
2.262	2.263	31.0	22	19	1.721	1.720	00.4	20	17
		13.0							

[1] Experimental data were obtained on a Philips diffractometer with filtered Cu radiation. Calculated spacings (d_c) are for $a = 9.42$ and $c = 6.88$ Å.

ELLIOTT [253] have considered at great length in an attempt to locate the position of the proton. The theoretical possibilities, assuming an ordered structure with respect to protons, can be represented schematically as: →||←, ←||→, and →||→, where || represents m and → represents the direction of displacement of the electron density from each mirror. Under any circumstance the symmetry of m (at 1/4, 3/4) is destroyed for an ordered arrangement, whether or not the diffraction data are sufficiently critical to resolve this matter.

The structure of chlorapatite is similar to that of fluorapatite except that the much larger Cl atom is located at (0, 1/2 c), rather than (1/4, 3/4), and thereby acquires a coordination number of 6, rather than 3, with respect to Ca. Chlorapatite is not always truely hexagonal; monoclinic symmetry ($P2_1/a$) has been reported [97] with $a = 19.210$, $b = 6.785$, $c = 9.605$ Å, and $\beta = 120°$. PRENNER [192] has suggested that the symmetry is related to the ratio of the halogen atoms — presumably to the amount of ordering of them.

Obviously, with equal numbers of F and Cl atoms, ordering could not take place for short intervals without destroying the symmetry of $P6_3/m$ with respect to the m planes.

Discovery of chlorapatite with monoclinic symmetry and $\beta = 120°$ (at room temperature) is interesting from the viewpoint of the interpretation of powder diffraction photographs of carbonate apatites inasmuch as ($hk \cdot l$) and ($kh \cdot l$) reflections (hexagonal indices) would have identical interplanar spacings and would be indistinguishable by this method. An identical situation arises for a triclinic crystal with $\beta = 120°$ and $\alpha = \gamma = 90°$; there would be no "line splitting" so the powder diffraction effects might be virtually indistinguishable from those produced by a truly hexagonal crystal. Complex pseudosymmetry occurs for structures resembling lazulite (monoclinic with $\beta \sim 119°$). For the closely related lipscombite structures the symmetry is tetragonal [149].

Vague references to "channels" in the apatite structure [253] carry the implication that the anions (F, OH or O) are missing at sites (0, 0, 1/4; 0, 0, 3/4) and state: "... the structure would not fall apart if all of these ions were removed (provided, of course, that the charge balance could somehow be maintained)." This speculation comprises a most dramatic violation of PAULING's electrostatic valency principle and has no straightforward evidence to support it among naturally-occurring apatites. Indeed, the density of apatite (3.1 to 3.2) does not suggest the presence of "openings" or "channels" of significant size inasmuch as the density is slightly greater than that of whitlockite [$Ca_3(PO_4)_2$]. Further discussion of this topic is presented in connection with crystal-chemical consideration of oxyapatite (voelckerite).

The most puzzling aspect of the structure is its manner of accommodation of carbonate ions as substitutions for phosphate ions, involving substitution of essentially planar groups for tetrahedral groups. A model presented by MCCONNELL [138] in 1938 and slightly modified in 1952 [140] suggests that there is no significant change in the positions of the oxygen atoms and that the change merely involves 4 carbon atoms interchanged for 3 phosphorus atoms: in effect 3 PO_4 become 4 CO_3 of the new structure. A well-ordered structure of this type would require the absence of a Ca atom in the site (1/3, 2/3, 0), presumably replaced by OH, H_2O or H_3O. Figures 9 and 10 indicate this hypothetical structure, and indicate 3 CO_3 groups parallel with the c axis and a fourth group perpendicular to c. The arrangement is symmetrical with respect to the 3-fold axis at (1/3, 2/3) of the diagram as well as the mirror plane at 1/4 c, provided three of the CO_3 groups remain parallel with c[1].

It is difficult to imagine an ordered structure containing coordinated groups of this magnitude without the consequences of a superstructure, of which there is no clear evidence. Furthermore, there is optical evidence [215, 147] that the CO_3

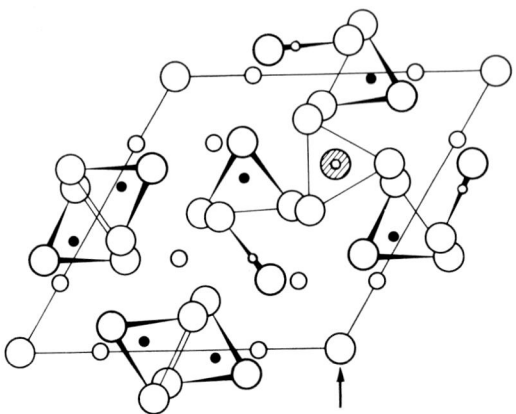

Fig. 9. Diagrammatic representation of $3 PO_4 \rightarrow 4 CO_3$. Small open circles are C atoms and large shaded circles are H_3O^+ at $(0, \frac{1}{2} c)$. The arrow indicates the origin which is the viewpoint for Fig. 10. (McConnell [140])

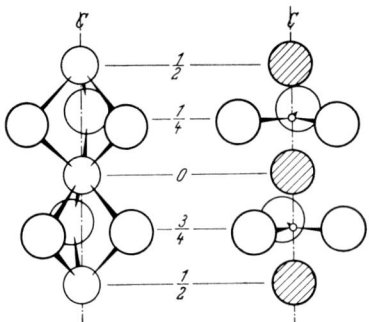

Fig. 10. Arrangement of oxygen atoms adjacent to 3-fold axis. Left: Ca atoms at $(0, \frac{1}{2} c)$ and O atoms at $(\frac{1}{4}, \frac{3}{4})$. Right: Displaced Ca atoms have in their sites H_3O^+ ions (shaded) and C atoms form CO_3 groups. (McConnell [140])

groups are not strictly parallel with $[c]$. The assumption, however, that the CO_3 groups are not parallel with the c axis carries with it the corollary that the m plane becomes inoperative inasmuch as C atoms could not lie in m, and other problems of coordination would arise were $3 PO_4$ groups not to become $4 CO_3$

---

[1] Attribution of the proposal for $3 PO_4 \rightarrow 4 CO_3$ to Hendricks & Hill [91] is incorrect. These authors did not correctly interpret the work of McConnell in 1938 [Fig. 9] which did not indicate that carbon substituted for calcium atoms but that carbon "displaced" Ca to form CO_3 groups, in which case the Ca sites $(1/3, 2/3, 0)$ were assumed to be vacant. In 1952 [140] it was suggested that these vacated sites for Ca probably contain OH^-, H_2O or H_3O^+, as is shown in Fig. 10. The work by Hendricks & Hill, furthermore, made no attempt to reconcile the sites for the carbon atoms with 3-fold symmetry of the structure.

groups [147]. Thus the postulation persists that 3 PO_4 become 4 CO_3 groups and the charges can be balanced locally by substitution of H_3O^+ for Ca^{++}.

It is probably futile to explain the carbonate apatites on the basis of hexagonal symmetry in view of the fact that most of these substances yield optical evidence of lower symmetry. For example, the crystals from Magnet Cove, Arkansas, display twinning that reduces their symmetry to monoclinic or lower, although this evidence was not presented by McConnell & Gruner [163] or by Carlström [32]. More recent examination by the author of Carlström's "single crystals" indicates that extinction is not rigorously parallel with the c axis and that they consist of sectors some of which show additive retardation and some of which show substractive retardation when the ε (actually α) vibration direction is very close to c and the first-order red plate is inserted. Similar twinning of lamellae parallel with the hexagonal prism surely reduces the symmetry to monoclinic, whereas the nonparallel extinction with c can only be interpreted as triclinic. The angular departures, if any, from 90° and 120° must be extremely small, however, because they have not been demonstrated by x-ray diffraction, and polysynthetic twinning often is not observable in hexagonal sections. Thus, for empirical reasons it is questionable whether a single crystal of carbonate apatite has ever been examined by any technique, and for theoretical reasons it is questionable whether such a crystal could be truly hexagonal.

Symmetry changes of the sort suggested by both carbonate apatites and by hydroxyapatite could result in the presence of polar axes and thus give rise to pyroelectric or piezoelectric properties which have been suspected to exist for bone. This topic will receive further consideration in connection with bone.

Isotypes of chlorapatite containing Pb, rather than Ca, have been investigated, and a direct structural analysis for vanadinite $[Pb_{10}(VO_4)_6Cl_2]$ by Trotter & Barnes [218] confirmed the predicted [92] close relationship to mimetite $[Pb_{10}(AsO_4)_6Cl_2]$.

Certain details concerning the structure of various apatite compositions have been attacked by methods involving vibrational spectra, and various interpretations have been applied to absorption bands within the infrared frequency range. Levitt & Condrate [130], as well as Klee [112], have shown, however, that some of these interpretations are erroneous and that Raman spectral information is essential to the correct interpretations of the infrared bands. Furthermore, it is pointed out that analysis by the site-group method is inadequate and that the factor-group method is required in order to take into account the interactions of the six phosphate groups within each unit cell.

Previously Adler [1] had indicated that both splitting and frequency changes were to be expected as a consequence of substitution of PO_4 groups for AsO_4 groups of mimetite.

Extension of the concepts of Levitt & Condrate to the carbonate apatites surely suggests that most of the interpretations which have been presented previously not only have a dubious basis, but some of them are essentially erroneous.

A recent study [187a] attributes the "long prismatic habit of natural apatite" to "pure screw dislocations which align with the c axis and mixed dislocations which deviate slightly from alignment with the c axis." The less frequent tabular habit remains unexplained.

5. Crystal Chemistry

More than 30 years ago [138] it was stated: "The structure of apatite seems to be remarkably stable, permitting a number of rather unusual types of substitution and involving a considerable number of ions." Crystallochemical investigations during subsequent years have extended the list of isomorphic substituents both qualitatively and quantitatively. However, it should be indicated that one new fundamental concept resulted from the investigations: Crystal chemical considerations do *not* require the balancing of electrostatic charges in such a way that a carbonate apatite can be written as $x\text{Ca}_3(\text{PO}_4)_2 \cdot y\text{CaCO}_3$, but must involve summation of the charges of all cations (Ca, P and C) versus those of the correct number of oxygens and fluorines which the structure will accommodate. Stating this proposition slightly differently, a carbonate apatite is not a mixture in any sense of tricalcium phosphate and calcium carbonate, but an individual structure $[A_{10}(ZO_4)_6X_2]$ with which its composition must be reconciled in respect to such physical properties as density and refractive index.

The application of such principles led not only to a clarification of the relationship between wilkeite and apatite and the discovery of ellestadite, but also to a demonstration for francolite that a CO_3 group cannot substitute for two F ions, although according to earlier chemical considerations supposedly this might be the situation. To be more specific, it was demonstrated that atomic weight units of a CO_3 group (ca. 60) are greater than for two F (ca. 38), and the density would necessarily increase if the volume remained the same or decreased. It was observed that the density decreased despite the fact that the volume also decreased, thereby eliminating the earlier interpretation, $2\text{ F} \rightarrow \text{CO}_3$. To be sure, it was further indicated that the comparatively large CO_3 group also would require an increase in the a dimension, whereas this was the very dimension which significantly decreased with occurrence of carbonate groups in the structure of francolite [81].

These details are mentioned merely to emphasize the unsatisfactory status of crystallochemical concepts with respect to apatite prior to 1937. As will become evident during following discussions, the last word on the crystal chemistry of apatite has not been written and may not be feasible in the near future. The situation is not improved by those who seek interpretations which are not consistent with recognized crystallochemical principles, or by interpretations that are based on spurious data, such as measurements made on a mixture of two phases, or on a single solid phase for which the chemical composition is merely surmised.

In the sections that follow the newer ionic radii of SHANNON & PREWITT [201] are used in preference to those calculated by AHRENS [2], in general, but inasmuch as most of the crystallochemical investigations antedate publication of the newer radii, it is realistic to examine various types of correlations on the same basis as that of the original proposals. Parenthetical Roman numerals refer to coordination numbers (CN) and frequently more than one radius will be given when the CN is not precisely known for the site under consideration. For example, the four Ca atoms located on the 3-fold axis cannot be said to have the same CN as the other six Ca atoms of the unit cell.

In unambiguous cases valence is not discussed nor is a charge necessarily indicated for the cations, so that the symbol for an element is interpreted to mean the atom with its normal charge. Likewise, the charge for anionic groups, such as PO_4 and CO_3, is not indicated except under special circumstances.

With respect to the "apatite group" — here denoting primarily calcium phosphates having a particular structure — isomorphic substitutions will be discussed with respect to substitutions for (i) Ca, (ii) P, (iii) F, and (iv) complex polyions.

5.1. Substitutions for Calcium

Varietal mineral names had been assigned to substances which subsequently were recognized as having the apatite structure. Dehrnite and lewistonite, as examples, were described in 1930 [122] and the chemical data were inadequate to permit their classification as apatites, although the optical properties were sufficiently close to arouse suspicion [123]. Both minerals produced x-ray powder diffraction patterns attributable to apatite [138] and both contained K and Na, as well as considerable water and, in the case of dehrnite, certainly CO_2. Thus, if there is a real difference between these varieties, and if there is justification for retaining these names, dehrnite probably applies to limited Ca → Na and lewistonite to Ca → K, without regard for the mechanisms needed to establish electrostatic balance.

Considering synthetic as well as natural substances, fairly complete substitutions for Ca have been demonstrated for Sr, Ba, Pb, Cd, Eu·· and Sn·· for halide and/or hydrophosphates. Complete substitution of Mn·· has been demonstrated merely for the chlorapatite structure [119], although manganapatites with as much as 10% of MnO occur naturally [138]. Chlorapatites containing Zn, Cu, Co, Ni and Mg, as partial substituents for Ca, have been synthesized from molten NaCl by KLEMENT & HASELBECK [117]. However, the tolerance of the structure for such atoms seemed to be limited to less than 5 Zn, 5 Cu, 4 Ni, 4 Mg or 4 Co atoms with respect to 10 Ca sites. The radii for these several divalent ions, according to SHANNON & PREWITT [201] and according to AHRENS [2], are compared in Table 5.1. Comparable ionic radii for monovalent ions are 1.33 (VII) or 1.02 (VI) for Na and 1.46 (VII) or 1.38 (VI) for K.

Table 5.1. *Comparisons of Radii of Possible Substituents in Apatite*

Coordination number	Radii in angstroms								
	Ni	Mg	Cu	Co	Zn	Ca	Sr	Pb	Ba
VII (VIII)	—	(.89)	—	—	—	1.07	1.21	(1.29)	1.39
VI	.700	.720	.73	.735	.745	1.00	1.16	1.18	1.36
AHRENS VI	.69	.66	.72	.72	.74	.99	1.12	1.20	1.34

When hydrothermal methods were used, hydroxyapatites containing 3 Cu or 2 Zn atoms per unit cell seemed to represent maximum limitations, suggesting that larger numbers of atoms with small radii are tolerated when the halogen ion has a larger radius (i.e., when Cl is present).

Empirical data on the changes in a and c periodicities are shown in Tables 5.2 and 5.3. Negative changes (i.e., contraction) of c are noted for chlorapatite, Cd-fluorapatite, and Cd and Mn chlorapatites when compared with ordinary fluorapatite. The radius of Mn$\cdot\cdot$ (0.820 [VI] or 0.93 [VIII]), being less than that of Cd (0.95 [VI], 1.00 [VII] or 1.07 [VIII]), places the Mn-Fap variant far outside the compositional area (Fig. 11) of "hexagonal apatites" as defined by KREIDLER & HUMMEL [119], although within the area for Mn-Cl-ap (Fig. 12).

Table 5.2. *Unit-Cell Dimensions **a** for Several Apatites (in Å)*
(Δ Represents Difference per Atom)

	F	Δ	OH	Δ	Cl	Δ (Cl–F)
Ca	9.37	.025	9.42	.105	9.63	.130
Δ	.034		.034		.024	
Sr	9.71	.025	9.76	.055	9.87	.080
Δ	.004		.012		.012	
Pb	9.75[a]	.065	9.88	.055	9.99	.120
Δ	.041		.031		.027	
Ba	10.16	.015	10.19	.035	10.26	.050
Ca	9.37		9.42		9.63	
Δ	−.007		−.009		.004	
Cd	9.30	.015	9.33	.170	9.67	.185
Δ					−.013	
Mn$\cdot\cdot$	—		—		9.54	

[a] An unpublished value obtained by GRISAFE & HUMMEL is 9.777 Å.

Table 5.3. *Unit-Cell Dimensions **c** for Several Apatites (in Å)*
(Δ Represents Difference per Atom)

	F	Δ	OH	Δ	Cl	Δ (Cl–F)
Ca	6.88	~0	6.88	−.050	6.78	−.050
Δ	.040		.040		.041	
Sr	7.28	~0	7.28	−.045	7.19	−.045
Δ	.002		.013		.015	
Pb	7.30	.055	7.41[a]	−.035	7.34	.020
Δ	.039		.029		.031	
Ba	7.69	.005	7.70	−.025	7.65	−.020
Ca	6.88	~0	6.88	−.050	6.78	
Δ	−.025		−.024		−.028	
Cd	6.63	.005	6.64	−.070	6.50	−.065
Δ					−.030	
Mn$\cdot\cdot$	—		—		6.20	

[a] Value given by BHATNAGAR [18]; it seems more probable than the value ($c = 7.429$) given by ENGEL [55].

The composition $[Cd_{10}(PO_4)_6F_2]$ is exceptional in showing contraction of both a and c, possibly suggesting that Cd is the smallest cation for which the end member with the fluorapatite structure is stable. Several apatite structures containing Cd have been studied by ENGEL [54], but the agreement is not good for a and c measurements obtained by KREIDLER & HUMMEL.

Substitution of Sn·· for Ca has been reported for the chlorapatite structure [117] and to a somewhat greater extent for a variant of hydroxyapatite $[Sn_9Ca(PO_4)_6(OH, Cl, F)_2]$ produced by hydrothermal synthesis [161]. For the latter preparation, however, $a = 9.45$ and $c = 6.89$ Å do not tend to confirm the radius of Sn·· (1.22 [VIII]) of SHANNON & PREWITT because this radius is nearly as large as that of Sr (1.25 [VIII]) for which the hydroxy variant has $a = 9.78$ and $c = 7.29$ Å [38], and the fluor variant has $a = 9.71$ and $c = 7.28$ Å [119]. Inasmuch as AHRENS' radius for Sn·· is smaller (0.93 [VI]) than his radius for Ca (0.99 [VI]) a question might arise concerning the relative radii of Sn··, Sr and Ca in apatites.

Yttrium and cerium and other rare earths have been reported for several natural varieties; so-called saamite, for example, contains 3 to 5% of rare earths in addition to SrO. Belovite is a variety far more complex quantitatively and qualitatively, and includes 3.60% of Na_2O. Abukumalite contains significant amounts of thorium, as well as yttrium, but also contains tetrahedral cations in addition to P, both Si and Al. Natural yttrium-containing apatite from Japan [183] contained 10.65% Y_2O_3 and had $a = 9.40$ and $c = 6.86$ Å. The question of the valences of Y, Ce and other rare earths in apatites is raised by the synthesis of $Eu_{10}(PO_4)_6(OH)_2$ [173] inasmuch as this composition gave $a = 9.73$ and $c = 7.22$ Å, comparable with $Sr_{10}(PO_4)_6(OH)_2$ with $a = 9.76$ and $c = 7.28$ Å. The radii [201] are given as 1.25 (VIII) for both Sr and Eu··, but as 1.07 (VIII) for Eu···. The slightly smaller periodicities for $Eu_{10}(PO_4)_6(OH)_2$ imply that Eu·· has a radius slightly smaller than Sr or that a small number of Na ions may have occupied sites (in addition to larger amounts of Eu) inasmuch as the preparation involved solutions of $EuSO_4$ and Na_3PO_4 and both Na and SO_4 ions can enter the apatite structure without great difficulty.

Thus far, the discussion has been concerned with cations involving coordination numbers of six or greater. A most interesting situation is the substitution of Al for both Ca and P. Through heating of morinite, FISHER & McCONNELL [62] found that something intermediate between $Ca_6Al_4(PO_4)_4(AlO_4)_2F_2$ and $Ca_8Al_2(PO_4)_5(AlO_4)F_2$ formed with respect to its Al content. The presence of 7.9% of Na_2O somewhat reduced the Al in the Ca sites below the two-to-one ratio, and it is not known whether the Na is essential to the structure's stability. However, it is noteworthy that the powder diffraction pattern was virtually indistinguishable from that of common apatites, both with respect to interplanar spacings and intensities, and thus gave no suggestion of a change of space group or significant change of a or c periodicities. The appreciable differences in ionic radii [201] with coordination number are indicated in Table 5.4. Whereas Al (IV) is slightly greater than twice that of P, Al (VI) is only slightly above half that of Ca (VI). Although this might be supposed to create an ordering among the Al atoms of the structure, no evidence in support of this supposition was apparent [62].

In summary, the substituents for Ca seem to be those with $CN > 6$ and valences from 1 to 3, although end-member compositions are not known for the F analogue for radii smaller than approximately 1 Å nor larger than approximately 1.4 Å. The possibility of substitution of H_3O^+ or H_2O for Ca ions will be considered under carbonate apatites.

Table 5.4. *Comparisons of Ionic Radii for Different Coordination Numbers*

CN	Ca	Al	P
VIII	1.12	—	—
VII	1.07	—	—
VI	1.00	0.53	—
IV	—	0.39	0.17

5.2. Substitutions for Phosphorus

Mimetite and vanadinite, which are Pb-Cl analogues involving AsO_4 and VO_4, respectively, have been known for many years; however, they belong to the subgroup of pyromorphite [$Pb_{10}(PO_4)_6Cl_2$] which will not be included as falling within the scope of the present work. They are frequently deeply colored, have high densities (>6.5), and are associated with other ore minerals in the oxidized zone of lead deposits. Varieties such as achrematite, bellite, and endlichite are not considered here [79].

Some varieties that are more closely related to apatite but which, nevertheless, do show substitution for P, are indicated in Table 5.5, without regard to their mode of occurrence insofar as the principal larger cation is not Pb.

Table 5.5. *Principal Cations of Natural Apatite Varieties*

Name	Large cations	Tetrahedral cations						
		S	Si	As	V	Al	C[1]	P
sulfate apatite	Ca	+						+
wilkeite	Ca	+	+					+
ellestadite[2]	Ca	+	+					
beckelite	Ca, Ce *et al.*		+					
lessingite	Ca, Ce *et al.*		+					
britholite	Ca, Ce *et al.*		+					+
abukumalite[3]	Y *et al.*		+			+		+
svabite	Ca			+				+
fermorite	Ca, Sr			+				+
saamite	Ca, Sr, RE							+
strontiapatite	Sr, Ca *et al.*							+
francolite[1]	Ca						+	+
dahllite[1]	Ca						+	+

[1] Carbon almost certainly does not occur with CN = 4 in the apatite structure. The distinction between francolite and dahllite is based upon the fluorine content (see Carbonate Apatites).

[2] Includes hydroxylellestadite [86].

[3] The Commission on New Minerals and Mineral Names of the International Mineralogical Association has approved the substitution of britholite-(Y) for abukumalite [63a].

5.2. Substitutions for Phosphorus

Table 5.5 contains only one example in which a significant amount of Al appears to occur with tetrahedral coordination and, although the synthesis of Al-rich apatite has been mentioned previously, the amounts of Al_2O_3 usually encountered in analyses of apatites often may be correctly assumed to be attributable to extraneous mineral contaminants (feldspar, clay minerals, etc.).

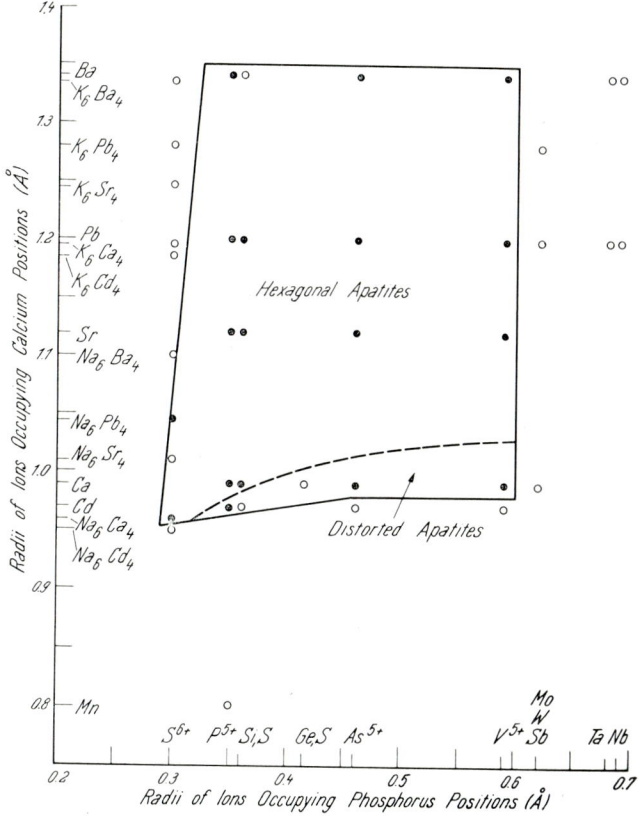

Fig. 11. Stability range for synthetic fluorapatite. Solid circles represent apatite structures whereas open circles represent other phases. (KREIDLER & HUMMEL [119])

While vanadium probably can substitute for P to a limited extent, it is seldom encountered in greater abundance than 0.05% (elemental V) in phosphorites, where it seems to reach a maximum of about 0.1% for samples from the Phosphoria Formation [82]. In francolite from Staffel, Germany, however, V_2O_5 is reported as 0.24% [80]. The radius of V^{5+} (0.355 [IV]) is identical with that of As^{5+}, so its insignificant occurrence may be related to lack of availability of the element. In phosphorites the meager amount of V is frequently greater than that of As, a matter which will be discussed under phosphorites.

In synthetic isotypes Cr probably is pentavalent [10, 61a] rather than tri- and hexavalent, as was assumed to be the situation in earlier investigations [176] which resulted in small, brilliant, green crystals of the Cl analogue. For earlier syntheses it was assumed that the Cr-apatite was anhydrous and showed a Cr : Ca ratio less than 0.6. For CN = 4 the radii are Cr^{4+} 0.44, Cr^{5+} 0.350 and Cr^{6+} 0.30, there being none given for Cr^{3+}. However, Cr^{3+} (VI) is given as 0.615, suggesting that Cr would have a radius > 0.50 were it to occur in the trivalent state with tetrahedral coordination [201].

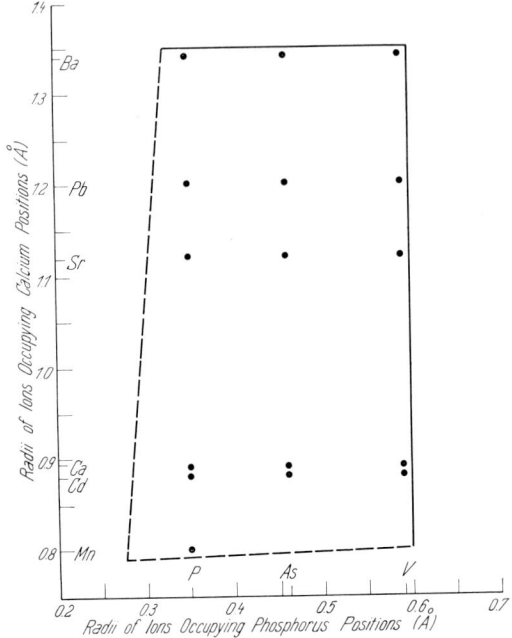

Fig. 12. Stability range for synthetic chlorapatites. Compare with Fig. 11. (KREIDLER & HUMMEL [119])

COCKBAIN [37] separates apatites (including analogues of vanadinite, mimetite and pyromorphites) into three categories based on the ratios of the radii of A and X ions of $A_{10}(XO_4)_6Z_2$. As in the work of KREIDLER & HUMMEL [119], COCKBAIN apparently used the radii of AHRENS [2], making adjustments for coordination numbers. However, it should be indicated again that there is a dependency upon the size of the halogen or hydroxyl (or oxygen) ion which is not taken into account by COCKBAIN. KREIDLER & HUMMEL recognized this relation to the extent that they give a field for chlorapatites, as distinct from fluorapatites (see Figs. 11 and 12).

Comprehensive coverage of the synthetic analogues of fluor-, chlor-, and hydroxyapatites is beyond the scope of this work — particularly inasmuch as many of these substances are more closely related to pyromorphite, mimetite and vanadinite. Nevertheless, it is interesting to compare the radii [201] of certain cations which might occur in apatites as substitutes for phosphorus (i.e., with CN = 4), such as

B^{3+}	Si^{4+}	Ge^{4+}	S^{6+}	Se^{6+}	Mn^{6+}	W^{6+}	U^{6+}
.12	.26	.40	.12	.29	.27	.41	.48 (Å)

With the exception of boron and sulfur, these cations are significantly larger than P (0.17 [IV]), but they are smaller than Al (0.39 [IV]) except for germanium, tungsten and uranium. Under natural conditions it is questionable whether Ge or W would enter apatite in amounts greater than traces except under extraordinary conditions of paragenesis. Uranium is discussed further under phosphorites.

In addition to the synthetic boron-containing apatite [178], the metamict substance called spencite fuses on heating to 1000° in air to produce glass and an apatitic phase that is believed to have the composition $Y_3(Ce, Pr, Th)Ca(Si_2B)O_{12}O$ [104].

Evidence from synthetic preparations [207], one natural occurrence [152] and numerous phosphorites [158] recommend the conclusion that the apatite will accommodate "tetrahedral hydroxyl" ions (groups of four OH ions as H_4O_4) analogous to their occurrence in hydrogrossular. In the latter case the end member $[Ca_3Al_2(D_4O_4)_3]$ has been investigated by neutron diffraction by FOREMAN [66] and the structure was found to be essentially similar to grossular except that the tetrahedral oxygen atoms were not centered by Si and each oxygen had an associated deuteron contributing to a DO group. The H_4O_4 group, to be sure, is considerably larger than the SiO_4 group, and in grossular to hydrogrossular this represents a change in a from 11.851 to 12.576 Å.

5.3. Substitutions for Fluorine

It becomes difficult to discuss the substituents X (of $A_{10}(ZO_4)_6X_2$) per se as a separate entity because of the complex interrelations with other types of substitutions, particularly substitutions of F and OH which may occur in sites other than (0, 0, 1/4) and/or (0, 0, 1/2). Chlorine, which occupies the latter coordinate position, can be replaced by other, even larger, halogen ions for pyromorphite. WONDRATSCHEK [245] has synthesized the analogues of Pb phosphate, arsenate, and vanadate with $X =$ Br. The vanadate with $X =$ I was also obtained, but the arsenate was not hexagonal and the phosphate was not obtained. The respective radii (CN = 6) for F, Cl, Br and I are [2]: 1.33, 1.81, 1.96 and 2.20 Å. It is to be remembered, however, that in apatite F has CN = 3, so its comparison with other halogens is not strictly valid.

The most perplexing question impinges on the possibility of vacancies among sites at (0, 0, 1/4). Not only has it been assumed that half of these sites may be unoccupied to yield $X_2 =$ O [oxygen] but it has been assumed that compounds of the type $A_{10}(ZO_4)$ can have the apatite structure. The argument is most intricate and is unlikely to be resolved soon in terms of any analytical method that has been applied to date. The original supposition in considering the problem of the oxyapatites was comparatively naive: two monovalent anions are merely replaced by one divalent anion [197]. The structural implications are really somewhat more complex: either there is a statistical vacancy among half of the sites at (0, 0, 1/4) or there is some reason for assuming that ordering takes place — for some unexplained reason — and half of these sites are empty. The latter situation, of course, would require a change of space-group symmetry which has not been demonstrated.

Neither the statistical vacancy of approximately half of the sites (0, 0, 1/4) nor the ordering and reduction of equipoint sites to half of their numbers seems to represent a highly probable explanation, and it should be pointed out that any such magnitude of missing large anions has not been recognized in oxygenated structures of this type. A specific case, however, involves a synthetic silicate pyromorphite which ITO [102] supposed to contain Pb^{4+} in order to obtain adequate cationic charges. While MERKER et al. [175] may have demonstrated that no tetravalent Pb is present, the possibility of additional protons in the structure cannot be ignored, so it becomes a question of precise analytical methods for detecting the presence (or absence) of hydrogen in these structures. Further consideration will be given to this topic in connection with synthetic apatites.

In an attempt to explain the substitution of CO_3 groups for PO_4 groups, it was suggested that the manner of substitutions might be $PO_4 \rightarrow CO_3OH$ or CO_3F. However, were this substitution realistic, there should be a direct correlation between carbon and $[(OH + F + Cl) - 2]$. On the contrary, this ratio was found to range from 294 to less than 1 for a series of natural and synthetic apatites [158]. This topic, while not directly related to substitutions for fluorine, seems to be of significant crystallochemical importance with regard to the "excess" fluorine frequently obtained by chemical analysis of phosphorites, and further attention will be devoted to this matter.

Lack of evidence for replacement of 2 F by CO_3 has been mentioned under structure and will be discussed in greater detail under Carbonate Apatites.

It has been supposed that the compositional variants intermediate between chlorapatite and fluorapatite do not comprise a continuous range, and this supposition might seem reasonable because of the significant differences in size, as well as differences in coordinate positions, of these two halogen ions. In contrast, however, is an apatite found in a marble near Matale, Ceylon [40], which contains F : Cl : OH : : 1.0 : 1.17 : 1.36 for a total of 1.975 anions, as compared with the theoretical 2. This same analysis shows 0.27 atoms of carbon when all cations (except H) are summed to 16. Unfortunately no diffraction data were reported for this material. A similar apatite from Uzunosawa, Saitama Prefecture, Japan, was found to have 0.75 F + 0.67 Cl + 0.76 (OH [86], for which the ratios are strikingly similar: F : Cl : OH : : 1.0 : 1.18 : 1.33. This apatite contained no carbon dioxide, however.

For fluorine, at least, the "tunnels" or channels", referred to by some authors, are surely tightly "plugged" inasmuch as the Ca to F bond of the triangular configuration (FCa_3) on the 6_3 axis ($0.25 \times 9.37 = 2.34$ Å) appears to involve distances that are slightly less than the sum of the radii for F (III) and Ca (VII), (1.30 + 1.07 = 2.37 Å). It is most unfortunate that a structural diagram was drawn in such a manner as to indicate the anion positions on 6_3 as vacant, creating the illusion of openings and thereby leading to numerous untenable speculations. Indeed, some authors have even attributed to apatite zeolitic characteristics and refer to such interactions as that of "bone char" with fluoride ions as though this reaction were — in every respect except reversibility! — comparable with that of a zeolite. The structural density of apatite shows no resemblance to that of a zeolite.

The questionable substitution of 2 F $\rightarrow CO_3$ has been discussed previously and will be discussed further under carbonate apatites. BONEL & MONTEL [20] claim to find large unit-cell dimensions for a, which they interpret to mean that the large group (CO_3) occurs on the 6_3 axis of "type A carbonate apatite". A similar product prepared by ELLIOTT [51] was examined by LEGEROS et al. [126] by x-ray diffraction methods. It is obviously a mixture of two or more phases [158]. Whether, indeed, such a compound as $Ca_{10}(PO_4)_6CO_3$ has been prepared in the laboratory, nothing of the sort has ever been shown to occur naturally, and a model indicating the coordinate positions of the atoms of the CO_3 group has not been described.

5.4. Complex Polyionic Substitutions

The degree of complexity of substitutions in apatite began to become apparent in 1937 with the discovery of ellestadite [$Ca_{10}(SiO_4)_3(SO_4)_3(OH,F,Cl)_2$] by MC-CONNELL [137]. Wilkeite had been discovered previously [50] but its relation to the silicate-sulfate end member had not been clarified beyond an indication that its optical properties permitted association with the apatite group. Wilkeite was known from only one locality and in terms of only one analysis; indeed, it was a mineral curiosity which had attracted little attention.

Before devoting attention to the extensive syntheses which came as an outgrowth of the relations demonstrated for ellestadite, another interesting aspect of the analysis should be mentioned: it yielded 0.61% carbon dioxide which could not be accounted for by contaminating carbonate minerals. The separation of the ellestadite from other mineral contaminants, to be sure, was admittedly imperfect, but no highly birefringent contaminant was observable. As in the case of francolite it became necessary to assign CO_3 groups to the structure of apatite [80, 140, 163].

This substitution of essentially triangular groups for tetrahedral groups ($PO_4 \rightarrow CO_3$) engendered extensive and somewhat controversial discussion, much of which was not presented by qualified structural chemists and was not enlightening. Nevertheless, the "problem of the carbonate apatites" is of crucial interest with respect to an understanding of the composition of teeth and bones and of phosphorites. A structural model for a carbonate apatite is discussed under structure.

Any discussion of carbonate apatites fairly quickly becomes involved also with the summation of halogen, hydroxyl and/or oxygen ions on the 6_3 symmetry axis because this summation can considerably exceed 2 in some instances, but this topic is reserved for more thorough consideration under Carbonate Apatites.

Soon after the original descripiton of ellestadite, KLEMENT [115] successfully synthesized $Ca_{10}Si_3S_3O_{24}F_2$ and, on the basis of the powder diffraction pattern, ascertained that it was an isotype of fluorapatite. Invesitgations of a large number of synthetic preparations have subsequently appeared by WONDRATSCHEK, KLEMENT, COCKBAIN, KREIDLER, HUMMEL, ITO, TRÖMEL, EITEL, BANKS, FRANCK, PRENER, and their several associates. The combinations that can result from one or more substituents for Ca and simultaneously one or more substituents for P are almost unlimited from the viewpoint of quantitative considerations, but many of these synthetic products are beyond the scope of present considerations and further reference to them will be reserved for specific illustrative examples.

However, the natural substance, caracolite, has been assigned the apatite structure and the formula $Na_6Pb_4(SO_4)_6Cl_2$ [200].

One of the earliest examples was prepared by KLEMENT [116]; it is $Ca_4Na_6(SO_4)_6F_2$, which seems plausible enough in terms of present-day knowledge of crystal-chemical principles, but which raises the question whether $Ca_4Na_6(SO_4)_6Cl_2$ also forms a stable apatite structure. The synthesis of britholite-abukumalite [217] should be mentioned, not only because of the relationship to natural substances, but because these substances represent "coupled diadochy" of A^{3+} and Z^{4+}, rather than A^+ and Z^{6+}, as shown by KLEMENT's work. Theoretically such fluor analogues would have for their respective end members the compositions $Ca_4Ce_6(SiO_4)_6F_2$ and $Ca_4Y_6(SiO_4)_6F_2$ but the natural substances are far more complex; britholite contains Na and P, whereas abukumalite contains Al in appreciable amounts.

Assignment of Al to Ca rather than P sites in the structure was feasible in terms of the fluor analogue obtained by heating morinite [62] because of the abundance of this element. Disregarding the Na content, which was not insignificant, however, would lead to a theoretical end member, $Al_{10}(AlO_4)_5(PO_4)F_2$, wherein twice as many Al atoms substitute for Ca as for P. In the synthetic substance, however, the approximate composition was:

for 10 A atoms $4.5\,Ca + 2.3\,Na + 3.2\,Al$, and
for 6 Z atoms $4.5\,P + 1.5\,Al$.

This may be near the upper limit of the structural tolerance for the same atom with two distinctly different coordination numbers. The radii involved were discussed previously.

There remains another possibility of even greater peculiarity, i.e., the possibility of Mn atoms, with entirely different radii and different valences, substituting in both A and Z sites of the structure. Substitution of $Mn^{\cdot\cdot}$ for A has been discussed, but the supposed existence of $Ba_{10}(MnO_4)_6(OH)_2$ [37] implies at least limited substitution of Mn with $CN = 4$, as has been discussed also in connection with the blue color of some apatites.

6. Synthetic Apatites: Applications

Previously it has been indicated that a carbonate fluorapatite (phosphorite) is the source of phosphoric acid, or in some cases elemental phosphorus. Some of the phosphorus is reconverted to apatite in its ultimate utilization, particularly as phosphors and as a scouring agent for dentifrices. To be sure, not all phosphors or scouring agents are apatites but these are important applications of synthetic apatites, and valuable information has been generated because of this economic importance.

6.1. Phosphors

Development of the fluorescent lamp provided impetus for the extensive investigation of phosphors which began about 1935. The earliest efficient conversion of electrical energy to ultraviolet radiation in a low-pressure mercury-vapor discharge tube and its further conversion to visible-range radiation through fluorescence, was not demonstrated by use of a phosphate phosphor, however, but through a zinc silicate comparable with willemite. The first white — actually a blending of several light colors — lamp utilizing a single-component phosphor depended upon proper ratios of Zn, Be, and Mn as a silicate.

The toxicity of beryllium-containing compounds required the search for a single-component white phosphor, and apatite was found to represent such a substance about 1949. Good control over color is obtained by varying the ratio of Cl : F as well as the Mn and Sb concentrations. As early as 1938 [84] it was recognized that the luminescence of apatite was attributable to the presence of manganese and rare earths. For natural apatites this conclusion has been confirmed [191].

Apatites, usually called halophosphates among the industrialists, have been subjected to detailed experimental survey with respect to the effects of Sb^{3+} and Mn^{2+} [185], such that it is now possible to predict departures from intended composition by using spectral measurements. Other substitutions for Ca, such as Sr and Ba have been investigated in conjunction with other activators, including cerium, copper and tin. A few elements (Al, Zr, Ce, La and Pb) were found by WACHTEL [233] to increase brightness. Oxidation of the activators Sb^{3+} and Mn^{2+} to higher valences should be avoided during preparation of phosphors which are essentially Ca apatites. WOLLENTIN [244] has investigated various substitutions for Cd in cadmium fluorapatites with respect to effects on color and brightness of the emission.

JOHNSON [105] has contributed a general summary on phosphors with "oxygen-dominated lattices", which includes general statements on theoretical mechanisms of energy transfer to produce luminescence, but such topics are beyond the scope of present considerations. Nevertheless, the extensive syntheses of apatites that have resulted from this application as phosphors must be considered in a general way. To be sure many of the compositions are quite removed from simple phosphates of calcium and involve considerable substitution for both anions and cations, as indicated in Table 6.1.

Table 6.1. *Synthetic Preparations with $A_{10}(ZO_4)_6X_2$ Structures Indicating Z for Various A and X Combinations*[1]

Large cation (A)	Anion(X)			
	F	Cl	(OH, O)	mixed
Ca_{10}	P, V, As, Si-S	P, V, As, Cr	P, Cr, Si-S, Cr-P	P, As
Ca_9Mg	P			
Ca_9Ni	P		P	
Ca_6Ce_4		Ge-P		
Ca_4Y_6			Si	
Ca_4Ce_6			Ge-Si	
Ca_4Nd_6			Si	
Ca_4Na_6	S			
Ca_5Cd_5	P			
$CaSn_9$				P
Ca_2Sn_8		P		
Sr_{10}	P, V, As	P, V, As, Cr	P, V, As, Cr	
Ba_{10}	P, V, As, Cr	P, V, As, Cr	P, Cr, V, Mn	
Pb_{10}	P, V, As, Si-S	P, V, As	P	V, As
Ba_2La_8			Ge	
Zn_2La_8			Si	
Zn_2Y_8			Si	
Cd_2Y_8			Si	
Ca_5La_5			Si	
Tl_2La_5			Si	
Pb_4Na_6	S			
Ce_{10}			Si	
Eu_{10}			P	
Mn_{10}		P		
Cd_{10}	P	P, As, V		
$(Ca, Na, Al)_{10}$	Al-P			

[1] Not included are a series of so-called "defect types" which show deficiencies of one or more cation and/or one or more anion per unit cell [58, 78].

Table 6.1 makes no pretense of being complete, but indicates compounds whose compositions have been reconciled with $A_{10}(ZO_4)_6(X)_2$ more or less precisely. $Pb_8Bi_2(SiO_4)_4(PO_4)_2$, for example, does not account for the X positions and therefore should be written $Pb_8Bi_2(SiO_4)_4(PO_4)_2\square_2$ in order to indicate their vacancy. A series of Sr-Nd silicates seems to indicate that occupancy of X positions is dependent upon the ratio of Sr/Nd [57]. Before discussing this topic, however, a few additional comments on the tabulation seem desirable.

It is noteworthy that both Al and Mn serve in dual rôles (they substitute for both A and Z), and while Al performs in this manner in the same structure, Mn seems to be predominately either Mn^{2+} or MnO_4^{3-}. Chromium had been assigned two valences, Cr^{3+} and Cr^{6+}, but BANKS et al. [10] have good reasons for believing that it is simply Cr^{5+}. The question of boron has not been adequately resolved [217], and Zn-containing phosphates require further investigation, although some silicates with the apatite structure have been reported [235a].

The tabulation is organized on the premise that the size of the X component is a primary factor as inferred by KREIDLER & HUMMEL [119] rather than the size ratio of $A : Z$ as inferred by COCKBAIN [37]. It seems that the chlorapatite analogue of Mn is known, for example, whereas that of fluorapatite is not. Many other vacancies in the anion columns should not be interpreted to mean that such compositions cannot be synthesized; probably many of them can.

Bromine and iodine can occur in apatites as principal X substituents, and the work of WONDRATSCHEK [245] should be consulted in this connection. Insofar as the author is aware such halophosphates have not been investigated for possible use as phosphors.

The question of the structure of apatite as $A_{10}(ZO_4)_6$ [actually $A_{10}(ZO_4)_6\square_2$] arises. Voelckerite has been mentioned previously. Absence of the centering of the triangular configuration of Ca ions (at elevation $1/4\,c$, adjacent to the 6_3 axis) by an anion, while possible, hardly seems probable when synthesis is brought about in an environment containing an abundance of anion-forming constituents, either in aqueous media or in melts.

6.2. Synthesis

One of the completely unjustifiable assumptions made in connection with syntheses is that the resulting product is a single phase with the apatite structure and contains the cations in exactly the same proportions as that of the original ingredients. Not only is this assumption frequently made for practical purposes, but it has even been made in connection with critical experiments which supposedly were intended to resolve theoretical questions. In other words, the synthetic product was not examined by microscopic and x-ray diffraction methods in order to ascertain the presence of a second phase (such as glass) and no chemical determinations were accomplished on a single apatitic phase. An outstanding example of such crude and faulty techniques is the supposed demonstration of the existence of an apatite with the composition $Ca_{10}(PO_4)_6(CO_3)$. In such an *anhydrous* composition no attempt was made to ascertain the presence or absence of water by direct methods despite the fact that the product presumably was obtained by precipitation from aqueous solutions. (The product was supposedly demonstrated to be anhydrous by incompetent interpretations of infrared absorption spectra, whereas a simple closed-tube test would have supplied far more valuable qualitative information in this instance.)

Numerous methods for preparation of synthetic apatites have been described in the literature, many of which involve significant concentrations of other ions (including Na and/or K), and are therefore objectionable. For production of hydroxyapatite LEHR et al. [127] recommended heating "a stoichiometric mixture of minus 200-mesh $Ca(H_2PO_4)_2 \cdot H_2O$ and $CaCO_3$ for 3 hours at 1200° C in an atmosphere of equal volumes of H_2O and N_2. Extract dried product twice with neutral ammonium citrate solution to remove calcium salts other than apatite. Wash product thoroughly with distilled water and dry at 100° C." Although this may yield a product low in carbonate, it is questionable whether a product entirely free from carbonate can be obtained by such a method unless the leaching with ammonium citrate is extraordinarily efficient.

SIMPSON [207] indicated that good, small crystals of hydroxyapatite can be obtained by slow release of Ca from EDTA to a phosphate-bearing solution at 100° C. It was found on analysis of his products that they invariably contained more than the theoretical amount of fluorine plus water (as OH) and that the ratios $(Ca + K + Na)/(P + C)$ ranged from 1.58 to 1.73, as compared with 1.67 for the correct stoichiometry. In view of SIMPSON's investigations it becomes highly questionable whether a synthetic product with the simple stoichiometry $Ca_{10}(PO_4)_6(OH)_2$ can be produced at 100° C (or less) in aqueous systems, and that most of the claims concerning such preparations are unjustifiable because the products were not subjected to valid analysis for water, carbon dioxide and alkalies.

Until comparatively recently, relatively little attention was given to the presence of carbon dioxide in starting materials, in alkalies used to control pH, or in the distilled water. As a consequence many preparations of so-called "pure" hydroxyapatite contained significant amounts of CO_3 ions, and a very simple qualitative test would have revealed the presence of carbon dioxide. Numerous conclusions based on such preparations are faulty as a consequence of assumptions concerning their compositions; some of these so-called hydroxyapatites were unquestionably mixtures of two or more phases.

Fluorapatite, according to LEHR et al. [127], can be prepared by heating "a stoichiometric mixture of β-$Ca_3(PO_4)_2$ and CaF_2 at 1370° C for 30 minutes in a current of dry N_2, with some CaF_2 upstream to minimize volatilization of fluorine (as CaF_2) from the reaction mixture; without this precaution the fluorapatite is low in fluorine and the refractive indices decrease to about $\omega = 1.627$ and $\varepsilon = 1.623$". Although these authors cite experiments of other persons involving "hydrolysis of octacalcium phosphate in boiling water containing dissolved fluoride", it is highly questionable whether $Ca_{10}(PO_4)_6F_2$ can be obtained by such a method [208].

In connection with synthesis of apatites in aqueous media — including hydrothermal processes involving bombs — it should be pointed out emphatically that it is virtually impossible to produce an apatite that does not contain carbon dioxide unless very deliberate methods for exclusion of carbon dioxide are used. Some investigators attempting to obtain $Ca_{10}(PO_4)_6(OH)_2$ have assumed that solutions of phosphoric acid and calcium hydroxide will produce an apatite free from carbon dioxide. However, $Ca(OH)_2$ will absorb atmospheric CO_2, and distilled water normally contains a small amount of dissolved CO_2. Indeed, the author has never discovered a reagent-grade calcium phosphate produced by any chemical vendor that would not produce effervescence on addition of a concentrated solution of HCl. (To be sure, the label on the bottle says nothing about carbonate contamination.) Furthermore, it should be indicated here that the "precipitated tricalcium phosphate" of most suppliers does not have Ca : P = 3 : 2 but is predominantly an apatite, containing more or less carbonate.

The literature prior to 1951 on synthesis of apatite has been compiled by JAFFE [103] but the abstracts contained therein bear virtually no relation to results obtained on natural materials, and the scant interpretations are not critical and include such erroneous generalizations as: "Carbonate-apatite apparently does not precipitate in aqueous systems." (This conclusion obviously denies either that bone is a carbonate apatite or that the physiological environment is an aqueous system, and thereby becomes an absurdity even in terms of reliable interpretations presented in the literature at least a decade earlier.) However, the compilation by JAFFE is useful as an indication of the quantity of haphazard experimentation, and it is of historical interest insofar as considerable attention is given to DALLEMAGNE's so-called tricalcium phosphate hydrate, a substance which has become as elusive as the unicorn of mythology.

6.3. Phase Relations

The phase relations of the binary system CaF_2—$Ca_3(PO_4)_2$ show a prominent eutectic at about 1200° C and about 60% CaF_2, but this diagram (repeated in numerous places, including [223]) does not include spodiosite. It is quite different from the diagram for $CaCl_2$—$Ca_3(PO_4)_2$, but again, chlorspodiosite is missing. For the latter compound (Ca_2PO_4Cl) the structure is now known [77]. The diagram for $CaCO_3$—$Ca_3(PO_4)_2$ includes a crystalline phase with the supposed composition $Ca_4CO_3(PO_4)_2$ coexisting with $CaCO_3$ below a eutectic temperature of about 1140°C [223]. Clearly, these systems require further investigation.

At pressures ranging from 500—4000 bars, WYLLIE [247] shows some rudimentary liquidus boundaries in the system CaO—CaF_2—P_2O_5—CO_2—H_2O. About the system CaO—P_2O_5—CO_2—H_2O at one kilobar he states: "The liquid at the eutectic coexists with calcite, portlandite, apatite (mainly hydroxylapatite, with possibly some carbonate-apatite of unknown composition in solid solution), and a vapor phase composed of $H_2O + CO_2$ which is very rich in H_2O."

No attempt will be made here to indicate what supposedly is known about the phase relations of the systems CaO—P_2O_5—CaF_2 or CaO—P_2O_5—$CaCl_2$ at various temperatures in "dry melts", "wet melts" or in aqueous systems for CaO—P_2O_5—H_2O. In aqueous media the establishment of equilibria may require extensive time, and at least one additional solid phase (brushite, $CaHPO_4 \cdot 2 H_2O$) frequently forms. Depending upon the pH and temperature, as well as impurities (including Mg and CO_2), whitlockite (β-$Ca_3(PO_4)_2$) is another possibility, as also is so-called octacalcium phosphate. Indeed, so confused are the interpretations that one author has produced the contradiction: "One cannot approach the same equilibrium for the two different directions — precipitation and dissolution." (Since equilibrium by definition is the steady state of concentrations, temperature and pressure where the velocity of dissolution is equal to that of precipitation, it must be concluded (i) that equilibrium was not obtained in his experiments because of inadequate time, or (ii) that dissolution $\rightleftharpoons$ precipitation is not strictly reversible — as is true of most silicates!)

6.4. Commercial Calcium Phosphate

The problem of use of a calcium phosphate as a scouring agent in toothpastes arises from the fact that these same dentifrices frequently also contain a fluoride, such as stannous fluoride. The affinity of hydroxyapatite for fluoride ions has been mentioned in connection with absorption of "bone char" for defluoridization of water supplies and in connection with the fossilization of bones. Thus, a difficulty arises in anticipating to what extent interaction between the fluoride component and the scouring agent will take place during any particular time interval, consequently affecting the "shelf life" of the product.

The fluoride, of course, is introduced for the purpose of reducing dental caries, and to be effective, fluoride ion concentrations should be within specific limits. The question of product development and control therefore impinges on rigorous reproducibility of the calcium phosphate. In view of the extreme complexity of systems containing Ca and P, and to some extent water, the number and compositions of phases present are difficult to control during manufacture. While the presence of hydroxyapatite is to be avoided insofar as possible, because of its reactivity with F, it has a fairly extensive stability range under conditions of fabrication of the scouring agent as well as after blending with other ingredients of the dentifrice.

7. Carbonate Apatites

Prior to a discussion of the inorganic substance of teeth and bones and/or discussion of phosphorites, it is essential to indicate the nature and extent of the complexities of these substances. In 1822 HAÜY [89] supposed that certain phosphorite minerals which occurred as vein fillings in Spain were to be considered as mixtures of a calcium phosphate mineral and a calcium carbonate mineral. In other words, HAÜY assumed that two mineral phases were present in the "crude calcium phosphates" at Estremadura, and that WERNER was in error in assigning to these mineral substances a separate name, phosphorite.

Indeed, the problem of the carbonate apatites was a very knotty one from the standpoint of the stoichiometry, but it was not feasible to account for the analytical results by assuming any ratio of admixture of apatite and a simple carbonate. To be sure, it could be assumed that electric neutrality obtained and various hypothetical phases, such as $Ca_3(PO_4)_2$, $CaCO_3$, CaF_2, $Ca(OH)_2$ and H_2O, could be summed in such a manner as to account for any particular analysis, in the same general way as one may attempt to account for the anion-cation balance of a water analysis. Such procedures continue even today, leading to such formulations as $Ca_3(PO_4)_6 \cdot CaCO_3 \cdot H_2O$. Not only is such a formulation irreconcilable with the structure of apatite but it has the Ca : P : C : H ratios 10 : 6 : 1 : 2, a situation which would be completely fortuitous were it to occur.

Only after some of the isomorphic substitutions in other minerals began to become clarified did it become possible to arrive at an understanding of the carbonate apatites. Through crystallochemical principles of general applicability, however, several puzzling features of the carbonate apatites have been resolved. For example, a plausible model has been proposed in order to account for the substitution of CO_3 groups for PO_4 groups (see Structure). This model has met certain physical and compositional tests which have been applied to it.

Chemical tests that must be met by any model are complicated by the difficulty of obtaining accurate ratios for the constituents, principally CaO, P_2O_5, CO_2, H_2O and F, several of which are volatile at various ranges of temperature, as is demonstrated by thermogravimetric analysis. Differential thermal analysis, likewise, has been applied in order to recognize energy changes, such as phase transformations, release of volatile components, interaction between phases, etc. However, the interpretations of DTA and TGA data are not completely straightforward for substances as complex as carbonate apatites.

7.1. Stoichiometry

Calcium to phosporus ratios have attracted considerable attention, as well they might, in connection with carbonate hydroxyapatites (*carHap*) and carbonate

7. Carbonate Apatites

fluorapatites (*carFap*). In general, these ratios exceed $10:6 = 1.67$ when Na, Mg, etc. are summed with the Ca and the divisor is solely P, and this fact was the principal chemical reason for assuming that the carbonate group substituted for the phosphate group. This type of substitution of a planar (CO_3) group for a tetrahedral group (ZO_4) does not appear to be restricted to the apatite structure, but occurs also for scawtite and ettringite, according to MCCONNELL & MURDOCK [165, 166].

The Ca : P ratio is not such a simple index as might be assumed initially, however. It has already been indicated that the substitution of CO_3 groups on the 3-fold axis [not the 6_3 axis] of the structural model must have concomitant effects on the number of sites available to Ca, there being no reasonable probability that Ca and C atoms could both be bonded to the same three oxygen atoms. The substitution of Ca, then, can be thought of as $Ca^{2+} \rightarrow H_3O^+$ or $Ca^{2+} \rightarrow H_2O$, so that the number of Ca atoms per unit cell might be less than ten even after addition of Na, Mg, etc. to give Ca' as $(Ca + Na + Mg + \ldots)$.

Water and fluorine are generally summed in any consideration of the crystal chemistry of either *carHap* or *carFap*. With small additions of chlorine, as may be necessary, the amount of $(F + OH)$ or $(F + OH + Cl)$ should be two such anions per unit cell. However, for many *carFap* compositions the amount of F alone may exceed two atoms per unit cell (or per 10 Ca' atoms) and there may be a significant additional amount of water that is not liberated below 300° C during any reasonable period of time. Strictly from the analytical viewpoint, then, any crystallochemical theory must be able to accommodate $(OH + F) > 2$.

Table 7.1. *Analysis of Apatitic Wood (Chemical Section, Illinois State Geological Survey)*

	Per cents of Oxides	Ratios (Atoms)	Cations = 16 (except H)	Cationic charges	Anionic charges
CaO	53.26	.9497	9.281	18.562	
SrO	0.04	.0004	.004	.008	
MgO	0.12	.0030	.029	.058	
MnO_2[1]	0.05	.0006	.006	.012	26 anionic
Na_2O	0.11	.0034	.033	.033	positions = 52
K_2O	0.63	.0134	.131	.131	
Fe_2O_3	1.45	.0182	.178	.534	
CO_2	5.26	.1195	1.168	4.672	
P_2O_5	37.54	.5290	5.170	25.850	
			16.000		
F	2.25	.1184	1.157		(1.157)
Cl	0.3	.0086	.084		(0.084)
H_2O	1.01	.1122	1.097	1.097	
			2.338	50.957	50.759

[1] Probably the Mn is divalent. Parentheses indicate negative values. Conclusions: The electrical charges balance within 0.2 electrons or 0.4 %, which is considered good agreement for calculations of this sort. There is an excess of 0.338 above requirements for the $(F + Cl + OH)$ positions; these hydroxyls are allocated to fill positions of displaced Ca atoms.

7.1. Stoichiometry

Results are given (Table 7.1) for a calculation on a carbonate fluorapatite based on a model having a total of 16 cations other than hydrogen. This calculation, which is based upon an analysis published by GLUSKOTER et al. [73], indicates summations for $A_{10}Z_6(O, OH, F)_{26}$ as follow:

$$10\ A \sim 9.28\ Ca + 0.38\ (Sr, Na, etc.) + 0.34\ (OH)$$
$$6\ Z \sim 5.17\ P + 0.88\ C$$
$$26\ (F, Cl, OH) = 1.16\ F + 0.08\ Cl + 1.10\ (OH) + 23.66\ O\ .$$

Again, about 0.3 atoms of carbon accompany substitution on the 3-fold axes to given CO_3—OH—CO_3, whereas the phosphorite analyses [195], subsequently mentioned (p. 55), indicate the more probable configuration CO_3—H_3O—CO_3. Particularly in low-fluorine francolites and dahllites, the analytical determinations for water tend to be somewhat below the theoretical requirement for $Ca \rightarrow H_3O$. The principal difference, however, between these analyses of francolites (immediately above and summarized under Phosphorites) remains in the indication of $PO_4 \rightarrow H_4O_4$ in one situation, but not in the other. Considering the tenacity of apatite of hydroxyl ions — which obtains at temperatures upward of 1000° C — such minor discrepancies are not surprising.

Obviously, water determinations "by difference" would have no validity for such complex compositions, and the tenacity of the apatite structure for F and OH is another complicating factor. For well-crystallized mineral specimens, however, adequate methods for determination of F are available, and the PENFIELD method, involving direct weighing of the condensed liquid, can be expected to yield as much water as can be liberated at the temperature attainable provided appropriate fluxing is accomplished [41]. After correction for any dissolved carbon dioxide, this water will still represent a minimal value, however, because of restrictions of the temperatures attainable. The possible incompleteness of recovery of water is related to the oxyapatite (voelckerite) problem [164].

Considering, now, the ratios $Ca : P : (OH + F + Cl)$, it has been well established that if Ca' is taken as 10, P can be considerably less than 6, whereas $(OH + F + Cl)$ may significantly exceed 2. This statement is true for *carFap* compositions which range from merely a trace of carbon dioxide to six or more per cent. In recent years an attempt has been made to assign the CO_3 groups to sites on the 6_3 axis, that is, to assume $2\ (OH + F + Cl) \rightarrow CO_3$ to give the overall composition $Ca_{10}(PO_4)_6CO_3$. The inherent simplicity of such a composition would be extremely nice, were there any justification for it from the viewpoint of either composition or structure. This composition can contain merely 4.27% by weight of CO_2, assuming that it contains no F or OH, so it is completely incompatible with almost all known compositions for natural substances, insofar as they show, with very few exceptions among *carFap*, $[(C/2) + F + OH] \gg 2$, and the exceptions probably are related to inadequate methods for F and/or H_2O. The structural incongruities have been mentioned.

From the chemical analyses emerge certain groups of data that must be reconciled with structural theory. If the $Ca + Na + Mg$ are taken as $Ca' = 10$, then: (i) there is a significant deficiency of phosphorus, and (ii) there is frequently an excess of $(OH + F + Cl)$ above 2. In 1970 McCONNELL [158] explained these analytical data in conjunction with what were called "hydrated carbonate apatites".

7.2. Structural Dimensions

Although the theoretical concepts presented by McConnell [158] are not complex, their utilization for the purpose of predicting the unit-cell dimensions (a) for these apatites was by no means a simple task. It was assumed, as a corollary to Vegard's law, that an additive relation should exist between the dimension a, several compositional functions and a constant, in the following relation (previously discussed under physical properties):

$$a_p = a_k + x(A) + y(B) + \ldots \qquad (1)$$

These relations seemed applicable to low-chlorine apatites, as a reasonable approximation, because the dimension (c) is not subject to pronounced change, except when chlorine is present.

This equation was then written in such form as to include hydrogen, carbon (as carbonate), fluorine, chlorine and sulfur (as sulfate), as follows:

$$a_p = a_k + x(H) - y(C) - z(F) + m(Cl) - n(S) \qquad (2)$$

Determination of the constant (a_k) and coefficients ($x \ldots n$) for the five compositional variables did not lend itself to straightforward computational methods, but it was possible to apply additional criteria and arrive at semiquantitative coefficients as follow:

Positive coefficients (a)	Negative coefficients (a)
hydrogen 0.0075	fluorine 0.015
chlorine 0.115	carbon 0.070

(expressed as angstroms per atom per unit cell)

The work of Kreidler & Hummel [119] suggests that the coefficient for S is extremely small, and McConnell [158] was unable to decide whether its influence on a was positive or negative, although n is shown with a minus sign in Equation 2. For the purpose of Equation 2 the constant of the structure was estimated as $a_k = 9.404$ Å.

The chemical data used in order to obtain these coefficients are given in Table 7.2. It should be mentioned that the coefficient for hydrogen must represent a mean value insofar as there are no criteria, beyond the calculated stoichiometric relations, for ascertaining how the protons enter the structure. It has been tentatively deduced that protons can enter such structures in several ways:

(i) the OH located at (0, 0, 1/4) can become H_2O,
(ii) a vacancy of a tetrahedral P site can give rise to H_4O_4, and/or
(iii) calcium can be replaced by H_3O.

Simpson [207 a], following the proposals of McConnell [145], has obtained evidence for such structural modifications and has suggested what influence these modifications might have on the unit-cell dimension a.

The case for $(ZO_4) \rightarrow (H_4O_4)$ is conclusive in connection with hydrogrossular since Foreman [66] has located the deuterons in $Ca_3Al_2(D_4O_4)_3$ by means of x-ray and neutron diffraction. While the extent of such substitution in apatites

7.2. Structural Dimensions

Table 7.2. *Comparisons of Measured and Predicted* **a** *Dimensions for Hydrated Apatites*

No.[1]	a in Å (meas.)	a in Å (calc.)	Atoms per unit cell					Key to Source[2]
			H	Cl	F	C	S	
1-f	9.29	9.341	1.32	0.01	2.17	0.59	0.07	W-M, 3
2-f	9.30	9.359	1.23	—	1.90	0.35	0.04	W-M, 7
3-f	9.31	9.358	1.02	—	2.18	0.30	0.02	W-M, 2
4-f	9.32	9.353	1.24	0.03	2.22	0.43	0.10	W-M, 1
5-f	9.34	9.323	0.52	none	1.98	0.79	none	SHM.
6-f	9.343	9.323	0.68	—	2.08	0.78	—	W-L, M 83
7-f	9.346	9.342	0.67	—	1.84	0.55	—	B-N, f
8-f	9.35	9.347	0.74	—	2.03	0.47	—	W-M, 5
9-f	9.356	9.349	0.84	0.06	1.00	0.76	—	H-T
10-f	9.36	9.337	2.33	—	2.17	0.73	0.20	W-M, 4
11-f	9.370	9.376	0.52	—	1.60	0.12	—	W-L, M 90
12-f	9.371	9.375	0.46	0.01	1.66	0.12	—	W-L, M 91
13-f	9.377	9.361	0.48	0.01	1.82	0.30	—	W-L, M 87
14-f	9.383	9.385	0.74	0.01	1.34	0.08	—	W-L, M 85
15-f	9.396	9.387	0.82	0.08	1.28	0.18	—	W-L, M 84
16-f	9.40	9.397	3.06	—	1.48	0.12	0.01	W-M, 6
17-d	9.419	9.358	1.59	—	0.22	0.79	—	B-N, d
18-h	9.420	9.426	4.48	—	—	0.17	—	S, K-3
19-h	9.427	9.447	6.80	—	—	0.12	—	S, K-2
20-h	9.432	9.440	4.94	—	—	0.01	—	S, K-1
21-d	9.454	9.405	4.00	0.12	0.02	0.61	—	M
22-ch	9.615	9.538	0.82	1.12	0.06	—	—	W-L, M 86
Idem	9.615	9.605	do.	[1.7]	do.	—	—	Idem

[1] Letters following the numbers are: f — francolite, d = dahllite, h — Hap and ch = chlorapatite.

[2] Sample numbers of original authors are given when appropriate; other symbols are: B-N is BROPHY & NASH [25]; f is francolite from Staffel; d is dahllite from Allendorf, Saxony, for which the oxide sum is 98.17%. H-T is HOFFMAN & TRDLIČKA [95], francolite from Kutná Hora. Sulfur and silicon were assumed to be elemental impurities by the authors. M is MCCONNELL [145] fossil dental enamel of mastodon, Ohio; sum of oxides is 99.39%. SHM is SANDELL, HEY & MCCONNELL [198], francolite from Tavistock, Devon. S is SIMPSON [207], K-1 to K-3 contained respectively 0.68, 0.84 and 0.97% K_2O. W-L is WALTERS & LUTH [235]. The water was "determined by difference". Their M 86 is reported to contain merely 1.12 atoms of Cl, but their c value suggests the amount should be 1.7 atoms, which would yield a predicted $a = 9.605$ Å. W-M is WHIPPO & MUROWCHICK [240]; all samples "recalculated to an impurity-free basis", including leaching with tri-ammonium citrate to remove $CaCO_3$.

still requires critical evaluation, it should be indicated that a small amount of this type of substitution seems fairly widespread and has become known as the "MCCONNELL-type defect" [22] in conjunction with theoretical structures for clay minerals. Several additional examples are listed by FOREMAN [66] but complete structural analysis has proven feasible only in connection with $Ca_3Al_2(D_4O_4)_3$, and here only because of the favorable ratio of D : O and the highest symmetry, space group $Ia3d$.

7.3. Refraction and Absorption

Although infrared spectral methods might seem to be applicable to location of CO_3 and OH versus H_2O versus H_3O versus H_4O_4 configurations within the structure, very little reliable information has appeared that is consistent with any other types of data or theories. Among such incompatible interpretations based on IR absorption bands are the following absurdities:

(i) the CO_3 groups are not present within the structure of francolite,
(ii) the OH groups of synthetic hydroxyapatite are absorbed on the surfaces of the crystallites which have extremely high surface areas, and
(iii) that a "degree of crystallinity" of bone can be ascertained and, when subtracted from 100%, will yield a percentage of "amorphous" inorganic substance present.

The difficulties of interpretation pointed out by LEVITT & CONDRATE [130] have already been mentioned. Empirically, however, some data obtained by LEGEROS et al. [125], CARLSTRÖM [32], and by EMERSON & FISCHER [53] strongly imply that the CO_3 groups must be present in more than one spatial environment, as would necessarily follow from MCCONNELL's hypothesis that one of four CO_3 groups is parallel with the basal plane whereas the other three are approximately perpendicular to the basal plane.

The orientation of planar CO_3 groups must be related to the negative birefringence which is several times greater for carbonate apatites and is accompanied by a small decline in the mean refractive index, according to MCCONNELL & GRUNER [163], CARLSTRÖM [32] and others. However, the extreme complexity of these structures makes accurate prediction of refractive indices from chemical analyses unlikely in the immediate future; for more simple compositions this topic is discussed under Physical Properties.

For a series of "fibrous francolites", CARLSTRÖM [32] obtained valuable quantitative data (Table 7.3). There is a consistent relation between the carbon dioxide content and the birefringence, as well as the *a* dimension of the unit cell. In general, the F content *increases* with an increase in carbon dioxide, whereas this is quite the opposite of what must occur if $2 F \rightarrow CO_3$. Unfortunately, the water contents of these same samples were not determined, but it is interesting

Table 7.3. *Correlations of Composition with Unit-Cell Dimensions and Birefringences*

Identification[1]	CO_2 (%)	F (%)	a (Å)	Birefringence
USNM 105532	4.39	4.59	9.334	0.0125
NRM 38745	4.22	4.06	9.342	0.0098
NRM 38648	3.18	3.83	9.349	0.0082
USNM 97339	2.57	3.94	9.350	0.0070
NRM 28258	2.01	3.83	9.352	0.0062
NRM 28260	1.32	3.71	9.360	0.0055
NRM 28259	1.26	3.78	9.361	0.0051
Syn *Fap*	–	3.78	9.371	0.0034

[1] USNM = U.S. National Museum. NRM = Swedish Museum of National History.

to note that the ignition losses (given by CARLSTRÖM, but not included in Table 7.3) for the second and fifth samples were about 7.2% and 2.7%, values which are significantly greater than the CO_2 contents but significantly less than the combined weights of fluorine and carbon dioxide. In the absence of SiO_2, the fluorine would not be highly volatile, but it is improper to assume that the water content represents the difference between loss on ignition and the CO_2 determination. CARLSTRÖM's data also indicated a more or less regular decrease in c from 6.895 to 6.884 Å accompanying the increase in a from 9.334 to 9.371 Å. However, changes in c are highly sensitive merely to Cl, and it is doubtful that a decline in c can be attributed to a decline in CO_2 *per se*. In this respect one notes that 1.26% of CO_2 produced a change in a of 0.010 Å, whereas no change in c is evident from CARLSTRÖM's data. For both NRM 28259 and synthetic fluorapatite he obtained $c = 6.884$ Å, despite a difference in birefringence of 0.0017.

A change in birefringence was one of the principal criteria that MCCONNELL & GRUNER [163] applied in conjunction with the hypothesis that the carbonate groups must be structurally incorporated within francolite. CARLSTRÖM [32] also attempted to consider these relations, but again, his calculations do not consider the effects of protons. Although there is some uncertainty about the precision of the ionic refractivities used in his calculations, one of his conclusions in interesting: "essentially all oxygen sites are occupied".

CARLSTRÖM's calculation of the molar refractivity shows a decline of 1.4 for francolite when compared with fluorapatite, despite a reduction of the volume. He attributes this decline to vacancy among Ca sites, and states: "In the material from Staffel which has a CO_2 content closely corresponding to one carbonate group per unit cell, the drop in R_M [molar refractivity] is only 1.4 instead of the expected 3.5." One deduces that he is concerned about 3 oxygen atoms per C versus 4 oxygens per P, inasmuch as such a postulation would account for the "expected 3.5." Were the situation one involving $2 F \rightarrow CO_3$, however, one would expect a considerable increase rather than a drop for the following reasons:

(i) taking the ionic refractivity of F as 2.2, that of oxygen as 3.5 and that of C as 0.4 (as assumed by CARLSTRÖM), then

(ii) for fluorapatite $R_M = K + 4.4$, and for francolite $R_M = K + 10.5 + 0.4$, giving not a "drop", as "expected", but a rise of 6.5. (K is the summation of ionic refractivities for $10\ Ca + 6\ P + 24\ O$ and becomes 106.4, using his values.)

Obviously the decline in the mean refractive index of francolite is completely incompatible with the substitution of CO_3 for 2 F, as was suggested in 1940 by MCCONNELL & GRUNER [163] by empirical methods, and probably is explicable only through detailed knowledge of such compositional changes as the substitution of F for O, rather than vacant sites in the structure.

Neither of these corollary propositions seems to be acceptable, as is evident from Table 7.4. $Ca'/(P' + C)$ ranges from 1.41 to 1.69, whereas $[(F + OH + Cl) - 2]/C$ ranges from -0.2 to 294. Only three of the latter ratios are below unity, a consequence forbidden by CO_3OH substitution, and this might be the result of incomplete recovery of water during analysis. On the other hand, departures in the direction greater than unity surely imply that if there is any correlation

Table 7.4. *Significant Ratios of Hydrated Carbonate Apatites*

Example[1]	Ca'/P'	$\frac{Ca'}{(P'+C)}$	$\frac{(F+OH+Cl\text{-}2)}{C}$	Example[1]	Ca'/P'	$\frac{Ca'}{(P'+C)}$	$\frac{(F+OH+Cl\text{-}2)}{C}$
1-f	1.82	1.64	2.5	10-f	1.87	1.64	3.4
2-f	1.75	1.65	3.2	16-f	1.71	1.67	21.2
3-f	1.73	1.64	4.0	17-d	1.60	1.41	−0.2
4-f	1.79	1.69	3.5	18-h	1.71	1.66	15.
5-f	1.80	1.57	0.6	19-h	1.62	1.58	40.
7-f	1.78	1.61	0.9	20-h	1.53	1.53	294.
8-f	1.78	1.64	1.6	21-d	1.69	1.53	3.5
9-f	1.87	1.67	0.1				

Ca' = Ca + Na + Mg + $\cdots$ (on the basis of atoms per unit cell). P' = P + S. For $Ca_{10}(PO_4)_6F_2$, Ca'/P' = 10/6 = 1.67.

[1] Examples carry the same references as those in Table 7.2.

between protons and C atoms it is a most obscure one indeed, and must involve atoms other than C displacing P.

Such deficiencies of P have been observed for a fairly pure natural substance from Sabinas Hidalgo, N. L., Mexico [152]; for several synthetic preparations by SIMPSON [207]; and for phosphorites. Where these deficiencies of PO_4 are greater than can be explained by 3 $PO_4 \to 4$ CO_3, it has been assumed that $PO_4 \to H_4O_4$ also occurs. Although this tentative conclusion might be questioned on the basis of the finely fibrous texture of francolite, it has recently been observed in conjunction with discrete single crystals of hydroxylellestadite that could not have relatively high surface areas and consequently could not adsorb H_2O [86].

The optical properties of a carbonate apatite are related to its specific gravity, of course, but in a more or less complex manner. Assuming hexagonal symmetry, the volume of the unit cell is $a^2 \cdot c \cdot \sin 60°$, so that one can calculate the molecular weight provided that the specific gravity is accurately known, or predict the specific gravity from the constituents of the chemical analysis (from the atomic weights). Unfortunately, the conditions for measurement of the specific gravity of a fibrous substance, such as *carFap* or *carHap*, are not ideal. Likewise, it is difficult to obtain an accurate evaluation of the molecular weight from a complex analysis. It is obvious, however, that $Ca_{10}(PO_4)_6CO_3$ would be greater by 22 a.w.u. than $Ca_{10}(PO_4)_6F_2$ since CO_3 is 60, whereas 2 F are 38 a.w.u. If then, the volumes are the same for these two compositions, the one with the carbonate group would have the greater specific gravity. All reliable observations indicate that this is not the case; despite its smaller volume *carFap* has a lower specific gravity. This is true for numerous natural substances.

7.4. Questionable Interpretations

In recent years it has been supposed that there are two different types of carbonate apatites: (1) in which CO_3 substitutes for PO_4, and (2) another in which

7.4. Questionable Interpretations

CO_3 substitutes for 2 F. With the exception of some very obscure hypotheses and the related calculations based on analyses of bone, all other evidences for $Ca_{10}(PO_4)CO_3$ are based upon the premise that the volume is greater, and that this expansion will permit incorporation of the large CO_3 group within the structure.

Some have supposed also that a dimensions as large as 9.518 kX comprise evidence [224] for 2 F $\rightarrow CO_3$. Unfortunately, the proponents of this hypothesis do not supply adequate information on their preparations. For example, the material reported by ELLIOTT [51], a sample of which was carefully examined by LeGeros et al. [126], produced at least five diffraction lines (within 2 θ from 28—35°) which are not attributable to the apatite structure [158]. Thus ELLIOTT's material is a mixture of two or more phases, rather than a single apatitic phase. Other experiments with synthesis of carbonate apatites are frequently so carelessly conducted as to defy critical appraisal from the amount of information supplied. Some authors, for example, do not even indicate whether Cl might be present in the apatitic phase, completely disregarding the pronounced increase in a that accompanies introduction of Cl. Chlorapatite has $a = 9.63$ Å, but $c = 6.77$ Å so the presence of Cl must be suspected when c is reported as being below about 6.87 Å.

There are numerous papers on *carHap* and *carFap* which are not cited here because (1) the conclusions contained therein are incompatible with accepted crystallochemical theory, (2) the experiments were not performed by appropriate methods, and/or (3) alternative interpretations of the data have been completely disregarded without satisfactory explanation. Some of the weakest of these reports deny the very existence of *carFap* and *carHap*, thereby reverting back to the state of knowledge in 1822 [89].

8. Phosphorites

Phosphorites are rocks composed essentially of carbonate apatites which are usually moderately high in fluorine and consequently are mineralogically francolite. From the standpoint of their economic utilization they must meet certain standards, but low-grade phosphate-containing rocks are widely distributed. Although commercial production in the U.S.A. has been confined largely to Florida, central Tennessee and a western district (including portions of Idaho, Montana, Wyoming and Utah), smaller deposits occur in Alabama, Arkansas, Georgia, Kentucky, North Carolina, Oklahoma, Pennsylvania, South Carolina and Texas. In the Soviet Union extensive phosphorite deposits are found in the Kara-tau area of southern Kazakh SSR.

With respect to the phosphate deposits of the western U.S., there is a bibliography of 89 pages [87], a biostratigraphic report on three Permian formations [249], and a transactions volume of the Geological Institute, Moscow, titled "Phosphoria Formation" [29]. Similarly, BUSHINSKY [28] has devoted 192 pages and 33 plates to "Old Phosphorites of Asia and their Genesis". These are a few examples indicative of a vast literature which is quite beyond the scope of the present work.

Phosphatic nodules have attracted attention from time to time with respect to possible models of formation; those of Agulhas Bank [South Africa] have been restudied by PARKER [186a]. D'ANGLEJAN [42a] has described a "bedded phosphorite facies" forming within recent sediments off the west coast of Baja California [Mexico]. In the latter investigation the structural carbonate of *carFap* of brachiopod shells served as a basis for age determinations ranging from 10,000 to 27,000 years. As is typical for marine phosphatic deposits, the phosphate-bearing substance was recognized as *carFap* [143, 195a, 197a].

BATES [12] recognizes three geological types of phosphorites: "(1) dark compact fine-grained strata, interbedded with mudstone or limestone, or (2) light-colored gravel beds, consisting of concretions, nodules, and pebbles of phosphatic material embedded in a sandy phosphatic matrix. A third type, residual from the first, is of lesser importance." With respect to the third type, weathering has been implied, but otherwise BATES has not indicated any paritcular mode of origin for either of his two principal categories. A noteworthy petrological study of phosphate rocks from various localities is that by TRUEMAN [219].

Some deposits of lesser economic importance do not fit BATES's classification, particularly those which result from interaction between bird excrement and insular rocks. Numerous deposits of this sort were described by HUTCHINSON [99] and certain details concerning the mineralogy have been supplied by FRONDEL [69] and by McCONNELL [139] among others. In general, calcium phosphates result from interactions involving calcareous rocks, including coral, whereas aluminum phosphates are usual when igneous rocks, including tuffs, are involved. (Nothing further need be mentioned about the aluminum phosphates since they are not apatites.) Such calcium phosphates are interesting because of their low fluorine contents. Mineralogically they are predominately dahllite, rather than francolite, but other less-stable calcium phosphates (monetite, whitlockite and brushite) coexist with the carbonate apatite, at least during the early developmental stages of such deposits.

Dahllite has an affinity for fluorine, of course, and probably is ultimately converted to francolite when an adequate supply of fluorine is available, given sufficient time. Such uptake of fluorine has been observed during fossilization of bones, and "bone char" or comparable substances have been used for reducing the fluorine content of water supplies, although other agents are frequently more efficient.

A most extensive literature exists on the probable mode of geologic origin of various phosphorite deposits of the world, but one topic persists as a fundamental question: Can phosphorite result as a direct precipitation product from sea water? Although this might seem to be a simple question and capable of a yes-or-no-answer, its complexity becomes considerably greater if one attempts first to answer the corollary question: What is phosphorite crystallochemically?

8.1. Relations with Sea Water

Experimental additions of several substances to sea water [14a], including chalk and silica, produced some interesting results. However, insofar as possible biologic influences are concerned, the degree of similitude with natural sedimentation is questionable [154].

There can be no question that some carbonate apatites are secondary. The conversion of wood tissue to apatite is unquestionable evidence of such secondary origin through replacement. On the other hand, the inarticulate brachiopod, *Lingula*, produces a shell which is francolite [150] so that precipitation from sea water (including related nutrient materials) by a biological agency must be admitted. Although simple inorganic chemistry may or may not be capable of yielding a theoretical answer satisfactory to everyone, it must be realized that sea water is by no means a simple inorganic solution. Indeed, sea water probably contains both catalytic and inhibitory agents that are far too complex to relate to simple calculations concerning the "saturation" of sea water with respect to a phosphorite of a particular assumed composition [154]. Ionic strength, i.e. salinity, may be an important factor, and magnesium ions seem to retard or impede the reaction when the Ca/Mg ratio approaches 4.5 to 5.2 (171).

With respect to fluorine, for example, SIMPSON [208] has indicated that an enrichment factor of 100 to 1000 times that of sea water would be required in order to produce fluorapatite at ordinary temperatures from aqueous solutions. His experiments stand in contrast with the ability of *Lingula* to accumulate fluorine within its shell of francolite, however, and this may be a further indication of the importance of organisms in controlling the equilibria of such reactions. Further discussion of the complex interrelations of bioprecipitation are discussed in the chapter on biologic apatites.

In any calculation of the solubility product for the system sea water and phosphorite, the most evasive unknowns are the types of ion complexing that occur, particularly in view of the multiplicity of anions. In addition to hydroxyl ions, the system is related to three phosphate ions, carbonate and bicarbonate ions, and fluoride and chloride ions. Despite the relatively high concentration of chloride ions in sea water, the tendency toward their incorporation within the solid phase is very small. A similar situation obtains with respect to sulfate ions although many phosphorites appear to contain a significant amount of sulfate [240] that presumably is present within the apatitic phase, rather than in a more soluble phase such as gypsum. Some of the other constituents obtained on analysis, such as silica, are more elusive with respect to assignment to the structure of apatite. For example, aluminum and silicon are known to enter the apatite structure, but a far commoner host for these elements would be a clay mineral. It must be remembered that phosphorites are rocks and usually contain at least traces of nonessential minerals, including quartz, opaline silica, pyrite, fluorite, hydrated oxides of iron and manganese, carbonates, and nonapatitic phosphates, as well as clay minerals and certain detrital minerals that are resistant to destruction by weathering processes.

CARPENTER [34] has discussed the marine geochemistry of fluorine and concluded that F is being removed from sea water about a hundred times faster than Cl, principally by incorporation within carbonates and phosphates of calcium.

8.2. Geochemistry: Enrichment

It is reasonable to conclude that the geochemistry of phosphorites is extremely complex, but not without significant interest insofar as their commercial utilization is concerned. Indeed, yttrium and the rare earths, which are potentially extractable during wet-process phosphoric acid production from phosphorites, have been estimated to exceed the amounts produced in the U.S. from bastnaesite and monazite ores [4]. Production of uranium, thorium, scandium and vanadium — the last in western deposits of the U.S. — is possible as byproducts, but the technology is not being applied on a significant scale.

Numerous investigators have been attracted to the topic of chemical enrichment of rarer elements in phosphorites from world-wide localities, and the average ranges indicated by SWAINE [213] are given in Table 8.1. As pointed out by TOOMS et al. [214], however, and as has been mentioned, the question arises whether these trace elements are actually incorporated within the carbonate apatite or are contained within the structures of admixed detrital minerals, authigenic minerals (either phosphatic or nonphosphatic) and/or organic matter. Nevertheless, it has been concluded [214] that "with few exceptions, it is strikingly apparent that relative to crustal abundance, the same group of elements is enriched in sea water as in phosphorite and those elements relatively depleted in phosphorites are similarly depleted in sea water. This illustrates that the characteristics of the depositional environment, sea water, must play a considerable part

8.2. Geochemistry: Enrichment

Table 8.1. *Comparisons of Abundance of Trace Elements in Phosphorites, Crustal Rocks, and Sea Water*

	I $\times 10^{-6}$	II $\times 10^{-6}$	III $\times 10^{-6}$	IV $\times 10^{-9}$
Ag	0.07	1–50	3	0.28
As	1.8	0.4–188	40[a]	2.6
B	10	3–33		4450.
Ba	425	1–1000	100	21.
Be	2.8	1–10		0.0006
Cd	0.2	1–10		0.11
Ce	60	9–85*		0.0013
Co	25	0.6–11.8		0.39
Cr	100	7–1600	1000	0.2
Cu	55.0	0.6–394	100	23.
Hg	0.08	10–1000*		0.15
I	0.5	0.15–280		64.
La	30	7–130*	300	0.0029
Li	20	1–10		170.
Mn	950	0–10,000	30	1.9
Mo	1.5	1–138	30	10.
Ni	75	1.9–30	100	6.6
Pb	13	0–100		0.03
Rb	90	0–100		120.
Sb	0.2	1–10	7[a]	0.33
Sc	22	10–50	10	<0.004
Se	0.05	1–9.8	10[a]	0.09
Sn	2	10–15		0.81
Sr	375.	1800–2000	1000	8100.
Th	7.2	5–100*		0.0015
Ti	4400.	100–3000		1.
U	1.8	8–1300	90[a]	3.3
V	135	20–500	300	1.9
Y	33	0–50	300	0.003
Zn	70	4–345	300	11.
Zr	165	10–500	30	0.026

I Abundance in crustal rocks, according to Mason [172].

II Averages computed by Tooms *et al.* [214] using data compiled by Swaine [213] for ranges in phosphorites of world wide distribution; * denotes total range.

III Modal values for the Phosphoria Formation according to Gulbrandsen [82], except *a* denotes average rather than mode.

IV Composition of sea water with 3.5% salinity according to Turekian [220].

in moulding the minor element characters of phosphorites." It must be understood, however, that the composition of the solid phase which forms in equilibrium with sea water is dependent upon some very complex energy relationships rather than simple functions of the ionic concentrations. Were this statement not valid, the apatite which forms should contain far more chlorine than it does in view of the comparative abundance of chloride ions in sea water. The relative abundance is greater than 10,000 in favor of chlorine over fluorine.

According to ALTSCHULER et al. [5] uranium may be as abundant as 0.1% in certain phosphorites because of secondary enrichment, and these authors indicate that tetravalent uranium probably enters the apatite structure as a replacement for calcium because of similarity of the ionic sizes, a suggestion that had been made a few years earlier [142]. PEACOCK & TAYLOR [187] recorded somewhat higher concentrations of U in phosphatized portions of the Carboniferous Limestone of Derbyshire and west Yorkshire, and MAZOR [173a] concluded that *carFap* is the "main mineral" in which P, F and U are "precipitated in marine sediments".

The results of KANESHIMA [108] enhance the conclusion that uranium accumulation is a secondary process; he found the phosphorite on the Ryukyu Island to be very low in uranium when compared with older, continental deposits. Further evidence of the uranium's association with the apatite is the fact that isolated fossil bones or phosphatic nodules may contain as much as 1% of uranium [5], an amount greatly in excess of that of the enclosing strata. A bibliography on U in phosphorites [83a] includes work as recent as 1968.

A general correlation between F and U contents, on the other hand, had been explained [203] by "chemisorption of the uranyl ion by two surface PO_4^{-3} groups to form a structure analogous to uranyl pyrophosphate". Such interpretations were not based upon contemporary knowledge of crystal chemistry, however, and coincide with an era during which colloid chemistry was called upon to explain many phenomena, including the presence of carbonate groups in apatite of bone. Were the U adsorbed on external surfaces of crystallites, there should be significant preferential liberation of U during leaching, but this was not found to be the case [101], and AMES [7] has found that U is not adsorbed but enters the structure of carbonate apatite. Under these circumstances, the supposed complexing of uranium and sulfate ions in sea water [83] does not seem essential to explain an inverse proportionality of these ions; merely the simultaneous substitution of $Ca^{2+} \rightarrow U^{4+}$ and $P^{5+} \rightarrow S^{6+}$ becomes improbable simply because of the increase in charges of the cations, regardless of the sizes of these ions.

The U occurs in both tetravalent and hexavalent states in phosphorites, and whereas it is not unreasonable to assume U^{4+} substitution for Ca in the structure of apatite, the occurrence of U^{6+} to form UO_4^{2-} groups might seem more perplexing. Nevertheless, the comparative radii do not eliminate this possibility. The tetrahedral radii, according to SHANNON & PREWITT [201], are (in Å):

P^{5+}	Si^{4+}	V^{5+}	Al^{3+}	U^{6+}	Fe^{3+}
0.17	0.26	0.355	0.39	0.48	0.49

To be sure, there has been no demonstration of tetrahedrally bonded ferric atoms (FeO_4^{5-} ions) in apatite, but significant amounts of AlO_4^{5-} are known to occur [62], and it is not unreasonable to presume that an extremely small number of UO_4^{2-} might occur also. The ratios of $U^{4+} : U^{6+}$ show very broad ranges, both for igneous apatites and for phosphorites, and any deductions based on such ratios (concerning the mode of occurrence of U in apatite) are necessarily somewhat speculative in view of the quantities involved.

Of the trace-element contents of fertilizers considered by SWAINE [213], evaluations of their desirability were given by him as indicated in Table 8.2.

8.2. Geochemistry: Enrichment

Table 8.2. *Desirability of Trace Elements in Fertilizers*

Of importance in agriculture			Significance uncertain			
necessary	harmful	doubtful				
B	As	Ba	Ag	Gd	Os	Ta
Co	Cr	Ga	Au	Ge	Pd	Te
Cu	Ni	Li	Be	Hf	Pr	Th
F	Pb	Ra	Bi	Hg	Pt	Ti
I		Rb	Cd	In	Re	Tl
Mn		Sc	Ce	Ir	Ru	W
Mo		Sr	Cs	La	Sb	Y
Se		U	Dy	Nb	Sm	Yb
Zn		V	Eu	Nd	Sn	Zr

It was admitted that the relationships are more qualitative than quantitative, however, and harmful amounts may be quite small for B, Cu, F, Mo and Se. SWAINE's appraisals seem to be more pertinent to plants than animals; some plants seem to thrive on traces of Se, for example, but accumulation levels of this element in certain plants are extremely toxic to sheep that ingest such forage.

Comparatively little is known about the behavior of individual elements during the postdiagenesis history of phosphorites. The amount of fluorine usually increases, and this qualitative observation has been used in an attempt to determine the age of mammalian bones. Even when the fluorine content is ascertained by direct methods, many variables necessarily exist, and age determinations based upon such indirect methods as x-ray diffraction cannot be considered reliable [148]. Similarly, the contents of chlorine and water (as OH) may be altered, and it must be remembered that the unit-cell contents of (OH + F + Cl) is not necessarily two [158]. A mastodon tooth, for example, showed 0.42% Cl within the dental enamel fraction [145] and this amount seems to represent an increase in chlorine during the fossilization process, if a judgment can be made on the basis of comparison with teeth of other living species of land mammals.

Strontium contents of fossil bones have been investigated [246], and the authors conclude it was an original component, whereas enrichment of the elements Si, Mn and Fe was found to represent fillings of the Haversian canals, and consequently post-mortem depositions [186] unrelated to chemical alteration of the apatite. Yttrium, on the other hand, is believed to enter the apatite structure during fossilization [186], as is the case of uranium, according to some authors [5].

Zinc, on the other hand, while comparatively abundant as a minor constituent of insular deposits investigated by KANESHIMA [108], was found to be removed during weathering by sea water and rain water. The amount of Zn in deposits of the Ryukyu Islands was found to be as great as 0.68%, whereas a maximum of 345 ppm is indicated in Table 8.1.

Mercury was not discussed by TOOMS *et al.* [214] but recent interest in accumulation of this element, supposedly resulting from pollution by industrial sources, justifies its brief consideration. Accumulation of Hg by certain animals, particularly fish, may account for the relative abundance of Hg in phosphorites (Table 8.1) and serves as another one of the many indicators that precipitation of phosphorites probably is not related solely to inorganic processes.

Indeed, the mechanism of incorporation of Hg in phosphorites is not known; it is possible that it is present as an organic compound. Furthermore, there is the unresolved question of its valence, although neither the mercuric nor mercurous ion probably is excludable from the apatite structure on the basis of the ionic radius. The fluorapatite which occurs at Cerro de Mercado, Durango, Mexico, contains merely 30—50 parts per billion (10^{-9}) according to a personal communication from E. J. YOUNG, and therefore approaches the limit of detection for the method of analysis used. Florida land pebble phosphate, according to COLLINS [39], shows a range from 10—1000 ppm.

Arsenic is known to enter the structure of apatite as a substitute for P, but this does not seem to be the manner of occurrence of As in Florida pebble phosphate [211]. Although there was no correlation between carbon and As for 51 commercial samples ranging from 3—17 ppm As, there was a relationship between As and Fe, and the light-colored pebbles showed a tendency toward smaller quantities of both iron and arsenic.

Conclusions concerning the depletion of Ce [64] in phosphorites of sedimentary marine origin are not borne out by the data in Table 8.1, but the degree of metasomatism may be an important factor, as has been suggested in the case of U [108].

Manganese, like As, is known to enter the structure of apatites in significant amounts, but it is doubtful that a significant amount of the Mn present in phosphorites (0—10,000 ppm) is incorporated within the structure because of the tendency of Mn to precipitate as complex hydrated oxides, usually also containing iron, in various sedimentary environments. Manganese is of geochemical interest, however, because the occurrence of manganese minerals may be the mechanism by which sequestering of other elements, particularly Ni, Co, Cu, Pb, Cr, Ba and Ti, takes place in phosphorites. Although crystallochemical considerations can exclude none of these elements, with the possible exception of Ti, from the apatite structure, nevertheless other host minerals are more likely, as is almost certainly the case for Al and Si — their occurrence in feldpars and/or clay minerals being most probable.

Resolution of the geochemistry of phosphorites, consequently, becomes a most complex problem with respect to identification of the mineral or organic host for any particular rare element. For this reason a correlation between organic carbon and some particular rare element may be misleading. Indeed, there may be partitioning of any given rare element among several minerals and/or organic compounds. Surely the assumptions that attempt to account for coupled substitution, such as Ce^{3+} and Na^+ for $2 Ca^{2+}$, in phosphorites approach absurdity in view of the quantities involved and in view of the multiplicity of other possibilities. It must be concluded, therefore, that our fundamental knowledge of the geochemistry of phosphorites is quite meager, and may not be amenable to improvement without quantitative knowledge of the several phases present in these rocks.

8.3. Carbonate Content

Leaching with a solution of ammonium citrate has been suggested [204] as a method for differentiating between admixed carbonate minerals, such as calcite, and the carbonate ions present in the structure of the carbonate apatite. While this procedure seems to be appropriate for producing a synthetic hydroxyapatite which is essentially devoid of carbonate ions, a few peculiar anomalies have arisen in connection with analysis of phosphorites. For example, if the ammonium citrate is quantitatively specific in its attack on simple carbonate phases, it becomes extremely difficult to explain an *increase* in the carbon dioxide content after leaching. Such a chemical method, of course, would be related to the surface areas of the substances (carbonate apatite vs. simple carbonate), as well as variations in their compositions, such as the magnesium content of calcite, for example. While x-ray diffraction is capable of detecting fairly small quantities of calcite or aragonite — less than 5% — this method does not seem to have been applied in a quantitative manner for practical purposes. The amount of simple carbonate present is of practical interest insofar as it bears on the amount of acid required for the production of superphosphate, for example.

On the other hand, the question of the CO_2 and water contents of phosphorites may contribute to the overall knowledge of carbonate apatites. KOLODNY & KAPLAN [118], for example, find the isotope enrichment of CO_2 from apatite is different from that of coexisting calcite. REEVES & SAADI [195] contributed analyses of eight "separated phosphate fractions" the calculations on which indicate [159]:

  (i) the substitution of C for P is not on a one-for-one basis,
  (ii) F substitutes for oxygen in francolites,
  (iii) the "excess" water substitutes for both Ca and P in the forms (H_3O) and (H_4O_4), and
  (iv) the model proposed by MCCONNELL [140], although fairly complex, appears to be applicable when the analytical data are adequate.

In contrast to these modern interpretations, for which there is now overwhelming evidence, stand reversions to some of the ideas expressed in 1822 [89]. It is unfortunate that present-day literature continues to be burdened by such archaic concepts, but it is even more unfortunate that the authors of such papers seem to believe that their proposals are novel in some sense.

9. Geology: Igneous and Metamorphic Occurrences

The characteristics of the phosphorites have already been considered briefly with respect to their crystal chemistry, geochemistry, petrography, and utilization. There is justification for their separate treatment from the viewpoints of geologists (particularly sedimentologists), physical chemists and chemical engineers. In the Phosphorites chapter little is indicated with respect to their geological age, geographical occurrences, or their modes of formation because these topics could, and probably should, be reserved for another author using some such title as "Rock Phosphates of the World".

Indeed, the diversity of occurrence among the various phosphate-containing strata is so great as to defy generalizations of value, and it probably can be assumed that most of the frustrations and polemics concerning such rocks arise from attempts to generalize. Whereas extensive and intensive geological investigation of two significant deposits might lead to a hypothesis for their formation, this generalization probably would not be appropriate for a third deposit.

Elsewhere consideration is given to the concept of formation of apatites at atmospheric and physiological temperatures and pressures. Indeed, the great masses of apatite have been discussed in other chapters, but the occurrences in igneous and metamorphic rocks cannot be neglected.

9.1. Igneous Rocks

Something approaching the composition of fluorapatite occurs throughout the entire compositional range of igneous rocks — ultrabasic to granitic — frequently as a minor accessory mineral. In calculation of the normative minerals, therefore, all analytically reported P_2O_5 is assigned to fluorapatite, whether or not adequate fluorine is reported. (The factor $P_2O_5 \rightarrow Fap$ is 2.37 on a weight basis; for Cl-ap it is 2.44.) The amount of apatite thus calculated as a normative mineral is nearly always less than 5% for undifferentiated rocks, and is usually within the range 0.1—1.0%.

KIND [110] considered the compositions of the apatites associated with a range of rock types:

	SiO_2 (wt. %)
I — Biotite granite from Lausitz, Germany	66.03
II — Monzonite from Ehrenberg near Ilmenau, Thüringen	60.13
III — Essexite, Rongstock, Bohemian *Mittelgebirge*	50.50
IV — Sodalite syenite, Schwaden, Bohemian *Mittelgebirge*	48.64
V — Ijolite, Jiwaara, Kuusamo, Finland	43.02
VI — Olivine nephelinite, Rossberg, Odenwald, Germany	40.39

9.1. Igneous Rocks

KIND separated the apatite from each of these rocks by density through the use of Clerici's solution [an extremely toxic aqueous solution of thallium salts of organic acids]. Compared with the normative amounts of apatite, calculated from the analyses of the rocks (quoted by KIND from other sources) his recoveries of isolated apatite cover a considerable range (Table 9.1). Such differences between the calculated (normative) *Fap* and the actual amount isolated through the use of a heavy liquid will depend upon the density range used for the apatite fraction, the amount and kinds of inclusions within the apatite crystals, and the quantities and compositions of other minerals that closely approach the apatite in density.

Some of the analytical results, such as 3.45% of water for the sodalite syenite (IV of Table 9.1), suggest that the rocks were not completely unaltered.

KIND gives the analyses for the several apatites isolated from the six rock types, as well as some of their physical properties. A few constituents reported are worthy of note (Table 9.2). He observed a correlation for TiO_2 and P_2O_5 when plotted

Table 9.1. *Comparisons of Normative and Recovered Apatite and Other Chemical Data for Different Rock Types of* KIND

Rock type	I	II	III	IV	V	VI
Wt. % P_2O_5	0.18	0.27	0.92	0.12	0.70	1.23
Normative	0.43	0.64	2.18	0.28	1.66	2.92[a]
Recovered	0.14	0.07	0.22	0.12	0.21	0.26
Yield %	33	11	10	43	13	9
TiO_2	0.69	0.45	1.91	1.88	0.63	1.12
H_2O^+	1.15	1.55	0.45	3.45	—	1.46
CO_2	—	0.19	—	1.42	—	1.66
SO_3	—	0.12	—	trace	—	0.60

[a] KIND's Table 1 gives 1.5% for some unexplained reason. Other differences are not significant.

Table 9.2. *Physical Properties Compared with Certain Chemical Components for Apatites from Different Rock Types*

	I	II	III	IV	V	VI
Color	weak yellow	dark gray	weak rose	clear gray	pure white	clear gray
Density	3.216	3.224	3.216	3.170	3.184	3.216
ω	1.641	1.657	1.641	1.659	1.641	1.637
ε	1.639	1.651	1.635	1.651	1.635	1.635
Δ	0.002	0.006	0.006	0.008	0.006	0.002
Cl	0.08	0.21	0.33	0.15	0.13	0.96
F	1.35	2.52	2.56	1.98	1.83	1.72
MgO	2.09	7.54[a]	1.42	2.57	1.06	1.38
MnO	0.15	n. d.	0.08	0.10	0.18	0.05
RE oxides	3.13	4.21	1.49	2.64	—	—
Fe, Al oxides	2.01	3.11	1.12	0.90	2.51	1.07

[a] The quantity of magnesia for apatite from the monzonite is surprisingly large, as is also the quantity of alumina plus iron oxides, and one might wonder whether ferromagnesian inclusions occurred in this apatite.

against SiO_2, a maximum occurring for each oxide within the range 40—50% SiO_2, with marked decline for more acidic rocks. For peridotite, both TiO_2 and P_2O_5 were very small quantities, resulting in an extreme slope between 38—42% SiO_2. One might suspect that his conclusion concerning ultrabasic rocks may be based on inadequate evidence. An augite porphyry from Colorado [239, p. 18] composed of augite, olivine, magnetite and glass contained 1.02% P_2O_5 and 1.35% TiO_2 but merely 37.83% silica; again, there is evidence of significant alteration because of 6.08% of CO_2 and 2.56% of $H_2O +$ and 1.24% of $H_2O -$. Thus, it seems hazardous to conclude, for example, that basic rocks (less than 40% SiO_2) are necessarily low in apatite content or that the apatite contained in such rocks is high in chlorine. Indeed, comparatively little is known about the compositions of the small apatite crystals which occur in most igneous rocks, because the more attention-taking occurrences are the ones for which analyses are available.

HOWIE [98] analyzed the apatite that has an abundance of about 5% in a pyroxene granulite from the island of Hitterö, southwest Norway. Expect for a composition about midway between Fap and Hap, it does not seem unusual. The atomic proportions, on the basis of 26 (O, OH, F, Cl), were calculated to be: 6.014 P; 0.012 Mg + .045 Fe^{2+} + 0.003 Mn + 0.012 Na + 0.003 Sr + 0.002 K + 9.887 Ca = 9.96; and 0.868 F + 1.098 OH + 0.056 Cl = 2.02. The unit-cell dimensions are $a = 9.382$ and $c = 6.887$ A; $\omega = 1.642$; density = 3.14. One notes that the minor constituents must produce compensating effects because the refractive index (ω) that would be predicted from the ratio of Fap to Hap is 1.6426.

Before leaving the magmatic rocks, it might not be out of place to add a precautionary note about inclusions within apatite crystals of metasomatized facies. It has been found through electron microprobe study that most of the RE (lanthanide series), U and Th occurs within monazite and xenotine inclusions of the apatite crystals, rather than in the apatite mineral itself, for some rocks at least [188], although primary igneous rocks may contain from 0.001 to 0.01% of U [5].

DEER, HOWIE & ZUSSMAN [46] assembled a dozen analyses of apatites, all of which are from occurrences other than undifferentiated igneous rocks. Three of the dozen are of carbonate apatites (contain more than 1% of CO_2), a topic to which these authors devoted considerable attention that is now of interest primarily in retrospect, because some of the concepts discussed by these authors subsequently have been completely disproven. Our attention will now be turned — as theirs was — to occurrences of apatite as segregations.

9.2. Segregations

Two well-known occurrences are those in Nelson County, Virginia, and the Kola Peninsula. KIND [110] supplied an analysis for apatite from the latter locality, and his results are compared with analyses obtained a few years later [231] in Table 9.3.

As noted (Table 9.3), normative nepheline has been deducted from VOLOD-CHENKOVA's analyses, apparently based on the assumption that Al_2O_3 is not a constituent of the apatite. This assumption has questionable merit [62], but under

Table 9.3. *Analyses of Apatites from the Kola Peninsula*

	Kind	Volodchenkova[1]	
		161	165
P_2O_5	40.84	41.00	40.62
SiO_2	—	0.68 [0.42]	0.26 [0.23]
RE_2O_3	0.77	0.93	0.98
Al_2O_3	0.68	[0.30]	[0.17]
Fe_2O_3	0.08	0.05	0.29
FeO		—	—
CaO	53.46	52.99	52.85
SrO	2.00	2.02	2.57
BaO	0.04	—	—
MnO	0.00	0.11	0.07
MgO	0.10	trace	0.01
Na_2O	1.21	0.52 [0.15]	1.10 [0.08]
K_2O	0.06	0.19 [0.06]	0.21 [0.03]
H_2O^+	0.19	—	—
F	1.10	2.60	1.79
Cl	0.31	—	—
	100.84	101.09	100.75
Less O	0.53	1.09	0.75
	100.31	100.00	100.00

[1] The analyses by Volodchenkova [231] were recalculated after deductions of amounts of oxides shown in brackets on the assumption that these amounts were present as a nepheline contamination.

any circumstance, all three analyses are low in fluorine suggesting either that there is an oxyapatite component or that significant OH and/or Cl has been missed. That is, for Kind's analysis:

$$H_2O+ \quad 0.19/1.79 = 0.11$$
$$F \quad 1.10/3.77 = 0.29$$
$$Cl \quad 0.31/5.66 = 0.05$$
$$\overline{0.45}$$

so this crude calculation indicates slightly less than half an ion (F + Cl + OH) instead of the ideal two ions. (The divisors are the amounts of H_2O+, F, and Cl for the theoretical compositions of *Hap*, *Fap* and *Cl-ap* without adjustment for the heavier and lighter substituents of the analysis.)

The three analyses are consistent in showing two (or more) per cent of strontia and less than one per cent of rare-earth oxides. It will be recalled that the variety saamite is from the Khibiny region; it contains SrO 11.42 and RE_2O_3 3.22%. Another high-strontia apatite has been reported from a syenite dike, cutting apatite pyroxenite, near Libby, Montana [124]. Rare earths, however, are not reported in the last analysis.

Within the past few years there has been extensive consideration by several Russian authors of petrologic relations involving apatite [232]. One of the more interesting studies of this series is the work of DELITSINA & MELENT'EV [47] which indicated no melting of the binary system apatite-nepheline below 1200° C, but for the system apatite-villiaumite a eutectic at 770° was detected for 43% apatite. Thus, the melting point was decreased about 850°[1]. A region of two liquid phases (aluminosilicate and fluoride-phosphate) was detected in an isothermal section of the ternary system from nepheline 0—85%, apatite 0—72% and NaF 15—100%. It was suggested that apatite-nepheline rocks formed by separation of a melt of phosphate-silicate liquid and that this model explains the deposits of the Khibiny (Kola Peninsula). Elsewhere it had been suggested that the apatite bands in magnetite ore bodies crystallized earlier than the bulk of the magnetite, and that when there was a large amount of apatite, it represented a differentiation to produce a residual melt of apatitic composition [72].

An interesting statement concerning energy relations has been made [210]: "The fugacity of phosphorus in equilibrium with apatite, a ubiquitous hydrous mineral, has been calculated for various mineral assemblages. The estimates, which are subject to considerable error, are lower for basanites and alkalibasalts than for tholeiites and range from approximately 10^{-14} at 1000° C to 10^{-16} bars at 750° C for fayalitic rhyolites." It's not clear how such thermodynamic data can be applied to metasomatic processes, however, and zonal variations in the compositions of carbonate apatites would seem to preclude such considerations.

Apatite associated with magnetite ore at the Mineville district, Essex County, New York, contains significant amounts of rare earths, and one sample was found to contain SiO_2 12.9, Ce_2O_3 12.9, RE_2O_3 23.8 and ThO_2 0.93 but merely 0.15% MnO. This apatite was metamict, in part, and presumably was not contaminated by monazite or bastnaesite [131].

9.3. Veins, Dikes and Sills

Dike rocks, of course, show a most diversified paragenesis, and the apatites which they contain show comparable diversity. While it is not possible to set forth any general statements, apatites containing manganese, strontium and rare elements are likely to be found in pegmatites and other types of dikes or sills. Indeed, FISHER [61] has supplied data on 70 species of phosphate minerals that are known to occur in pegmatites, either as a result of primary crystallization or as alteration products, so the paragenetic relations must be most complex. Thus, while apatite may be a primary mineral, it must be remembered that carbonate apatites can form at atmospheric temperatures.

Manganiferous varieties frequently contain 4 or 5% of MnO [194], but a very dark-colored specimen from central Finland contained 7.59% MnO [230] and another sample from a granitic pegmatite near Buckfield, Maine, contained $10.3 \pm 0.2\%$ MnO according to a partial analysis by GOLDICH [138]. For the later, a crude calculation indicates substitution of Mn for Ca merely to the extent of 1.5 of the 10 Ca ions, whereas complete replacement of Ca by Mn seems to be feasible through synthesis — but only for the Cl analogue.

---

[1] The melting point of apatite is about 1620° C.

9.3. Veins, Dikes and Sills

There is a dearth of good, recent analyses of apatites from veins and dikes; scrutiny of such analyses suggests that they tend to involve Cl and/or OH substitution for F, as well as substitutions of several elements for Ca. Thorough examination of material from a perthite-quartz pegmatite, cutting a granodiorite near Eagle, Colorado, indicated from 0.1 to 0.2% of oxides of Mn, Fe, Na, S and H; other prominent constituents determined by spectroscopic methods were 0.3 Si, 0.34 La, 0.69 Ce and 0.23 Nd. Undetected were: Ag, As, Au, B, Be, Bi, Cd, Co, Cr, Er, Eu, Ga, Ge, Hf, Hg, In, Li, Lu, Mo, Nb, Ni, Pb, Pd, Pt, Re, Sb, Sc, Ta, Tb, Te, Th, Ti, Tl, Tm, U, V, W, Zn and Zr [251]. This apatite is quite unspectacular when compared with some others, both in the types and quantities of substituents.

Apatite from the Morefield pegmatite, Amelia County, Virginia, has a bright orange-yellow fluorescence for the entire range of ultraviolet radiation, as well as bright yellow for x-rays [177a]. Persistent phosphorescence was observed for x-radiation, but not for UV, and strong thermoluminescence diminished fairly rapidly. The fluorescence presumably is caused by manganese and/or rare earths (chiefly cerium and yttrium) which are present in amounts from 0.5 to 1.0% (RE_2O_3) and 2—3% MnO.

Bukanov [27] has described apatite of tabular habit from "Alpine-type veins" that is an intermediate variant between *Fap* and *Hap*, and contains Na, K, Mn and Al substituting for Ca. The question of Al substitution in apatite remained obscure until it was found [62] that heating of the rare pegmatite mineral morinite produced an apatite in which 25 atomic per cent of Al substituted for both Ca and P in the approximate ratio of 2 : 1. Comparatively little is known about the occurrence of Al and Fe in natural apatites. Frequently these elements, as well as Mg, are assumed to be present as constituents of a contaminating encrustation or of inclusions, but abukumalite is supposed to differ from britholite primarily by the presence of AlO_4 groups in the former. Both minerals are silicate phosphates which were subsequently recognized as having the structure of apatite [212].

Substitution of strontium and rare earths is most pronounced in an "oxyapatite" from an alkaline-syenite pegmatite of the Burpala intrusion [190]; the amounts recorded are SrO 14.70 and RE_2O_3 13.03%. Other pegmatitic Sr-containing apatites have been mentioned above. The paragenesis of strontiapatite is complex [248]. In addition to 46.06% of SrO, this variety contains oxides of Ca, Ba, Mg, Na, Th, RE, Al and Si, and almost half of the F positions are occupied by OH. The refractive indices given are 1.651 for ω and 1.637 for ε; the difference (0.014) is unusually high for an apatite that does not contain carbon dioxide.

In conjunction with their studies of luminescence of apatite specimens, Portnov & Gorobets [191] have tabulated the RE_2O_3, Y_2O_3 and MnO contents for numerous localities. Their Table 1 is reproduced as Table 9.4.

9.4. Carbonatites

Apatite is the principal phosphorus-containing mineral in carbonatites, and in southern and eastern Africa such occurrences are becoming of increasing economic importance. Deans [44] has presented an excellent summary of several such deposits in which apatite may comprise 90% of certain ring dikes. He lists

Table 9.4. *Rare Earths and Manganese Contents of Apatites* (PORTNOV & GOROBETS)

	RE_2O_3 $+Y_2O_3$, wt. %	La	Ce	Pr	Nd	Sm
Kimberlite, Yakutia	0.88	8.0	63.0	1.0	19.8	3.0
Serpentinite, Shabry, Urals	0.15	8.0	24.0	—	23.8	—
Carbonatite, Kovdor[1]	0.70	15.4	45.6	8.9	19.0	1.0
Carbonatite, Vuori Yarvi	1.00	15.0	47.0	9.0	19.4	5.0
Carbonatite, Guli	0.60	13.0	43.0	9.5	17.5	9.8
Gabbroids, Kragero, Norway	1.00	6.4	36.0	25.0	4.0	5.0
Gabbroids, Kushva, Urals	1.50	18.0	47.0	16.0	12.0	1.5
Gabbroids, Kusa, Urals	0.11	11.0	38.0	—	34.0	—
Urtite, Khibiny	0.35	12.4	54.0	2.5	22.0	—
Alkalic pegmatite, Lovozero	4.30	33.5	48.0	5.0	11.0	1.5
Alkalic pegmatite, Burpala[1]	9.50	14.0	46.0	7.0	23.1	7.5
Fenite, Vishnevyye Mountains	3.00	23.0	45.4	12.2	12.7	2.9
Alkalic pegmatite, Il'meny	2.20	20.0	49.0	10.5	11.3	4.5
Granite pegmatite, generation I, Kazakhstan[1]	1.20	20.8	49.0	4.2	15.2	2.1
Granite pegmatite, generation II, Kazakhstan[1]	1.90	4.9	26.0	1.0	20.0	15.0
Granite pegmatite, Chupa Karelia	1.00	2.5	18.0	4.9	7.5	11.4
Granite pegmatite, Mama, E. Siberia	0.80	7.2	16.8	10.0	14.0	7.2
Granite pegmatite, E. Transbaikal	2.40	1.0	10.7	8.0	13.0	17.0
Granite pegmatite, Izumrudnyye mines, Urals[1]	0.70	13.0	36.0	6.5	22.0	1.5
Metasomatic bodies, Slyudyanka	0.12	10.5	41.6	—	24.5	—
Metasomatic bodies, Aldan	0.60	16.0	44.0	20.0	9.5	5.5
Alpine veins, polar Urals	0.02	—	—	—	—	—

[1] Mean of three analyses.

eight analyses which show significant ranges of constituents. Excluding four that are ore concentrates and contain other minerals, their compositions are given (in part) in the upper half of Table 9.5. It is noticeable that the material from Busumbu, Uganda [43] has almost exactly the ideal amount of F for *Fap* but contains additional amounts of water, equivalent to 30% of that required for *Hap* of H_2O+, and total water amounting to 46% of the ideal for *Hap*. Again, it seems unlikely that the 1.80% of CO_2 forms CO_3 groups which displace F or OH ions on the 6_3 axis of the structure, as has been inferred from the results of some extremely careless experimentation.

The three partial analyses (Table 9.5) of apatites from Kenya and Tanzania were supplied by P. PRINS [University of Stellenbosch] and have not been published previously. The Canadian specimen is from an occurrence believed to be a carbonatite in Faraday Township [82a]. An elaborate study of apatite-calcite relations at Oka, Quebec, Canada, by GIRAULT [72a] included investigation of the inclusions, but also revealed appreciably greater amounts of rare earths ($>2.5\%$ RE_2O_3).

9.4. Carbonatites

Eu	Gd	Dy	Ho	Er	Tu	Yb	Y	MnO. wt.%
1.0	0.8	—	—	1.0	—	0.1	2.3	0.028
6.0	4.0	—	—	2.0	—	1.0	31.0	0.005
1.2	1.6	—	—	3.0	0.2	0.3	3.8	0.010
0.8	1.0	—	—	0.4	0.1	0.1	2.2	0.010
1.6	1.0	—	—	1.0	0.2	1.0	2.4	0.010
1.0	1.4	14.0	—	1.0	0.1	0.1	1.0	—
0.6	0.6	—	—	0.3	—	0.1	4.0	0.010
—	5.0	—	—	—	—	1.0	11.0	0.120
1.7	1.5	—	—	1.5	0.3	0.3	2.8	0.050
0.2	0.1	—	—	0.1	—	—	0.6	0.150
0.5	1.5	3.0	0.2	0.1	—	0.2	3.5	0.010
0.3	0.6	—	—	0.3	0.1	0.1	2.5	0.200
0.3	0.4	—	—	0.3	0.1	0.1	3.5	0.150
0.1	1.0	—	0.3	0.1	—	0.2	7.0	0.150
0.5	4.5	4.5	0.8	0.1	—	0.7	22.5	0.200
—	5.4	15.0	1.0	0.1	—	0.2	34.0	0.540
1.2	6.7	12.0	—	1.9	—	0.1	22.9	0.100
—	14.0	4.0	0.8	0.1	—	0.4	31.0	0.330
0.5	3.0	2.0	—	1.5	—	1.0	13.0	0.850
—	5.2	—	—	5.2	1.0	1.5	10.5	0.100
0.6	0.4	—	—	0.4	0.1	0.1	3.4	0.100
—	34.0	—	—	—	—	9.0	57.0	0.005

Table 9.5. *Minor Constituents of Apatites Associated with Carbonatites*

Locality	Oxide							
	SrO	RE_2O_3	MgO	F	Cl	H_2O^+	CO_2	Others
Busumbu, Uganda	—	—	0.14	3.79	—	0.53	1.80	0.41[a]
Kivu, Congo	0.8	—	—	4.1	0.05	0.12	0.2	0.9[b]
Tanganyika	0.62	0.84	—	1.7	0.1	0.1	0.2	0.31[c]
Zambia	4.72	0.48	0.10	3.40	0.05	0.59	n. d.	0.45[d]
Rangwa, Kenya	0.52	0.18	0.21	3.10	n. d.	0.12	0.34	0.64[e]
Mbeya, Tanzania	0.65	0.34	0.45	2.84	n. d.	0.29	0.43	0.35[f]
Oldonyo Dili, Tanzania	0.54	0.19	0.16	2.46	n. d.	0.23	0.57	0.61[g]
Ontario, Canada	0.18	0.76	<0.01	4.0	—	<0.01	0.57	0.53[h]

Others: [a] H_2O^- 0.30 Fe_2O_3 0.11. [b] SiO_2 0.9.
[c] MnO 0.04 Na_2O 0.26 K_2O 0.01. [d] Fe_2O_3 0.45.
[e] Fe_2O_3 0.03 ZrO_2 0.09 H_2O^- 0.52. [f] Fe_2O_3 0.06 ZrO_2 0.11 H_2O^- 0.18.
[g] Fe_2O_3 0.17 ZrO_2 0.13 H_2O^- 0.31.
[h] MnO 0.15 Na_2O 0.18 K_2O 0.03 H_2O^- 0.03 Si 0.07 Fe 0.036 SO_3 <0.03.

9.5. Metamorphic Rocks

Some of the most peculiar compositions for apatites are encountered in connection with the metamorphic rocks of hydrothermal and/or pneumatolytic origin, or in low-temperature veins associated with various rocks (Fig. 13). These are likely to contain significant amounts of replacement of PO_4 groups by CO_3 groups within the structure. The original occurrence of dahllite [23] was a secondary

Fig. 13. Photomicrographs of hydrothermal apatite from Magnet Cove, Arkansas. Left: Thin section of crystals showing overgrowth of carbonate apatite at the ends of fluorapatite crystals; the polarizing prisms were almost in crossed position. Note that retardation is far superior for the carbonate-containing portions. These same portions do not have parallel extinction and are twinned. Right: Another type of carbonate-apatite overgrowth forming a lamellar structure about a hexagonal crystal. The individual lamella do not have parallel extinction and represent twins having a composition plane (10.0). (McConnell & Gruner [163])

deposition as crusts on earlier apatite crystals. Similar occurrences of carbonate apatites are known from Tonopah, Nevada; St. Paul's Rocks, Atlantic Ocean; Katzenbuckel, Odenwald, Germany; Richtersveld, Cape Province, South Africa [228]; the Khibiny region [48]; Kutná Hora, Czechoslovakia [95]; Peñascos de la Industria, Durango, Mexico [67]; François Lake, B.C., Canada [189]; and other localities. For some of the specimens studied more recently iron (FeO 1.12%) was believed to replace calcium [95], and Carlström [32] has indicated that the carbonate apatite from Magnet Cove, Arkansas [163] contains appreciable yttrium.

The francolite analyses show a range of carbon dioxide, but rarer constituents include merely SrO 0.13% [44] TiO_2 0.06% [100] and V_2O_5 0.24% [81], in addition to varying amounts of oxides of Mg, Na, K, Fe, Al, Si and S. They tend to show an excess of F above the theoretical requirement [182] and may also contain Cl and water.

A thorough study of the apatite occurring in iron ore at Cerro de Mercado, Durango, Mexico [252] shows RE_3O_3 1.43% (with (Ce, La, Pr, etc.)/(Y, Yb, Gd, etc.) having a ratio of 8.6. In this study, Young [252] recognizes two different types of iron ores associated with igneous rocks: (1) containing magnetite, martite and hematite in variable combination, and (2) containing titaniferous magnetite and/or ilmenite and occasionally rutile. He lists several examples of both types, but the grouping together of the Mexican and northern Swedish ores seems strange insofar as GEIJER [72] refers to the later as "magmatic", whereas FOSHAG [67] describes a paragenesis that includes calcite, diopside and sepiolite in the early stages but later formation of goethite, colorless apatite, dahllite, quartz, calcite and barite. It can probably be concluded that present knowledge of the geologic histories of some of these iron ores is as imperfect as is the knowledge of the compositions of the apatites associated with them. Some of these apatites are high in SiO_2 [131] but the yellow crystals from Durango are not. The age of the crystals from Durango has been estimated by etching of fission tracks to be 36.3 ± 3.5 million years [170].

Probably the complexity of the paragenesis of the metamorphic iron ores is only exceeded by that of the calcicsilicate contact deposits. Such a deposit at Crestmore, Riverside County, California, has proven to yield many new minerals, including ellestadite and wilkeite which have SiO_4 and SO_4 groups replacing PO_4 groups. Associated with the ellestadite were diopside, wollastonite, idocrase, monticellite, okenite and blue calcite [137]. A new variety, hydroxylellestadite, has been reported from the Doshinkubo ore deposit, Saitama Prefecture, Japan, in addition to wilkeite and an apatite containing prominent amounts of $F + Cl + OH$ [86]. The Japanese occurrences are similar to those at Crestmore, and the minerals (the sulfate-silicate apatites) differ principally in containing more CO_2, aside from their contents of P_2O_5, SiO_2 and SO_3. The minor constituents for the silicate-sulfate apatites are shown in Table 9.6. The apatite showed no CO_2, Na_2O 0.11, MnO 0.05 and SrO 0.55% in addition to a trace of MgO.

Table 9.6. *Minor Constituents of Sulfate-Silicate Apatites*

	CO_2	Na_2O	MgO	MnO	SrO	P_2O_5
Ellestadite	0.61	—	0.47	0.01	—	3.06
Hydroxylellestadite	1.65	0.34	trace	0.04	0.28	0.66
Wilkeite (Japan)	2.18	0.06	trace	0.05	0.86	15.90

A silicate apatite (SiO_2 12.9%) has been mentioned and sulfate apatites are known from several localities [275].

It should be indicated that the chemical compositions of apatites, particularly in situations where metasomatism is involved, may show significant variability. This situation is well illustrated by compositions of wilkeites at Crestmore, but appears to be applicable to many localities, including the *Hap* at Holly Springs [177] and to some extent to the apatite from Durango [67]. Thus, the identity of locality offers little assurance that specimens are similar in composition.

The hydrothermal apatite that occurs in a talc schist at Holly Springs, Cherokee County, Georgia, has attracted considerable attention because of its supposed relation to the composition of teeth and bones. However, it is a most inappropriate analogue insofar as no carbon dioxide is reported in the analysis by REYNOLDS

[177]. The water content (determined by loss on ignition, and after correction for volatilization of F) is reported as 1.86% which is somewhat in excess of the theoretical requirement (1.79%). A most noteworthy example of excessive water in apatite is revealed by analysis of material from Mercedes mine, Sabinas Hidalgo, Nuevo León, Mexico [30]. In this case the summation for (F+Cl+OH) positions of the structure is 3.2 rather than the theoretical 2 ions, and there is a significant deficiency of PO_4 [153]. The occurrence is described as white to yellowish crusts in cavities of dark-colored sedimentary rocks, in close association with whitlockite [30]. "No evidence of any high-temperature activity in the vicinity has been reported." Information on the composition and occurrence of material at Kemmelten (Kt. Uri) in Switzerland is inadequate.

The substances just mentioned (except the Swiss occurrence) are dahllites essentially, because of their low F contents. Another occurrence of dahllite is that of the sedimentary manganese ore deposits at Eplény, Hungary. GRASSELLY [76] deducted significant amounts of impurities (carbonates, opal, hydrated manganese oxides) from an analysis of a lens from the Upper Liassic deposits and obtained (on recalculation) 1.36% of CO_2, 0.46% F, and 1.01% H_2O+ as belonging to the apatite. Additional data obtained by infrared absorption and x-ray diffraction were used for characterization of the phosphorus-containing mineral of the ores at Eplény and at Úrkút. The latter is somewhat higher in both F and CO_2.

9.6. Lunar Rocks and Meteorites

Two analyses of lunar apatites are somewhat different with respect to minor constituents. The first analysis (Table 9.7) is by FUCHS [71], and a significant amount of the RE_2O_3 reported is Gd_2O_3, 0.66%. The rock (No. 10044) contains apatite as inclusions within a new mineral, pyroxferroite. Other minerals present included fayalite, silica minerals, native iron, and ilmenite.

The second analysis, by microprobe, is that given by ALBEE & CHODOS [3]. The apatite was contained within a fragment from a lunar soil that contained hypersthene, plagioclase, a glass rich in K_2O and BaO, and another phosphate mineral tentatively identified as whitlockite.

Again, a question concerning the contamination of the apatite crystals by inclusions might arise [129], but for apatite in rock No. 10044 the amount of RE_2O_3 is small when compared with apatite at Essex County, New York [131].

Meteorites of the stony variety may contain apatite as a minor accessory mineral; insofar as the composition is known, these crystals are the Cl-*ap* variety. So-called merrillite was formerly thought to be an apatite, but it is now believed to be whitlockite with some Na replacing Ca.

Table 9.7. *Analyses of Apatites from Lunar Rocks*

P_2O_5	CaO	FeO	SiO_2	F	Cl	Y_2O_3	RE_2O_3	Others
38.7	52.1	1.5	2.3	3.3	0.03	1.2	2.54	—
38.6	50.9	0.58	1.12	3.12	1.10	0.07	0.29	1.13[a]

[a] Na_2O 0.21, K_2O 0.24, MgO 0.30, BaO 0.03, Al_2O_3 0.23 and TiO_2 0.12%.

In this chapter no attempt has been made to cover all of the geological occurrences of apatite. Indeed most of the information contained in this chapter is taken from fairly recent sources. When combined with the data presented in the chapter on phosphorites, it becomes evident that the petrology suffers from inadequate information on petrography and mineralogy, and in turn, these suffer from inadequacies of analysis by chemical and spectrochemical methods. In some instances crystallochemical data are given, but attempts to ascertain the presence or absence of OH, F or Cl ions and CO_3 groups solely by infrared spectroscopy are questionable in view of the extreme complexity of interpretations by such indirect methods [157].

10. Biologic Apatites

Phosphates are essential to the existence of both plant and animal life. All vertebrates, and some invertebrates, are endowed with physiological mechanisms for accumulating both calcium and phosphorus to form skeletal tissues and teeth. Such metabolic processes are of major importance in connection with the egg-laying characteristics of birds, for example, but are related to vital activity of all vertebrates.

Plants, and most invertebrates, do not maintain a reservoir of calcium and phosphorus in inorganic form and therefore are dependent upon a continuous intake of these elements in order to sustain life. Aquatic plants assimilate phosphate from water, whereas terrestrial plants obtain phosphate from the soil. In turn, the animals acquire phosphates from plants, as well as other animals, and the phosphates are returned to water and soil after death of the animals. This recycling of phosphorus does not provide for uniform distribution or availability, and the ultimate reservoir is the oceans from which precipitation of phosphorites takes place.

From the viewpoint of applicability, one of the more important aspects of a theoretical knowledge of apatite is concerned with its relationship to teeth and bones. Nevertheless, the interpretations are very complex and the diversity of conditions under which the experiments have been performed has led to considerable confusion and many conflicting statements in the literature.

Under the topic of carbonate apatites, the reader has been prepared for a theory on the crystal structure of these substances, dahllite and francolite, but it must be remembered that the size of the individual crystallites of bone precludes the application of such straightforward methods as single-crystal analysis. Dental enamel and other biologic apatites are not greatly superior, but do offer some advantages, at least with respect to some analytical methods.

Much speculation, some of it far from useful, has resulted from consideration of the kinetics of aqueous solutions which approximate the composition of normal blood serum. Many such considerations include far-reaching assumptions that completely vitiate any probability of reasonable interpretation. For example, calculation of the solubility product for "bone salt" requires an accurate knowledge of the composition of the solid which is necessarily in equilibrium with the solution, but most authors have completely disregarded carbonate ions — present in both the solid and the solution. Furthermore, the effects of catalytic and inhibitory agents are usually disregarded completely, despite the fact that such substances are known to be of significance in most physiological processes. Related discussion involves the precipitation of phosphorite from sea water — a matter previously mentioned.

10.1. Teeth and Bones: Composition

Probably the most controversial topic in connection with analytical data on bone is that concerning the ratio of Ca ions to phosphate ions. For hydroxyapatite the ionic ratio is ideally $10/6 = 1.667$, but rarely is this true for bone. The dependence of this ratio upon other substituents — principally carbon atoms and protons — has been discussed under carbonate apatites, and while there has been considerable debate on whether the best analyses yield a value in excess of 1.667 or below this value, this argument is of little significance with respect to the theory to be discussed here. It has been shown (Table 7.4), that Ca'/P' may cover a range from 1.87 to 1.53 for hydrated carbonate apatites of mineral or synthetic origins [158]. Results on bone that are significantly below 1.667 — particularly those below 1.60 — almost certainly were obtained by unacceptable chemical methods.

The results of the diligent efforts of ARMSTRONG & SINGER [9] indicate that bovine cortical bone gives values slightly above the ideal ratio. Their analysis for dry, fat-free material is given in Table 10.1. When citric acid is added the summation is slightly below 50%. The percentages are for the elements (except CO_2 and citric acid), and can be converted to weight percentages of oxides, thus:

CaO	MgO	Na$_2$O	K$_2$O	SrO	P$_2$O$_5$	CO$_2$	F	Cl	Sum*
37.56	0.72	0.99	0.07	0.04	28.58	3.48	0.07	0.08	71.54

(* The sum has been corrected for F and Cl by -0.05.)

Adding the nitrogen and citric acid yields 77.32% of the material as accounted for. The remainder (22%) is undoubtedly organic carbon, chemically-combined water and such moisture as would remain present in the "dry" bone. There seems to be little probability of ascertaining the amount of chemically-combined water directly from bone in view of these uncertainties.

The analytical problem is most complex for bone insofar as an attempt is made to remove the organic portion without altering the composition of the inorganic

Table 10.1. *Analyses of Bovine Bone and Human Dental Enamel*

	Bone [9]	Enamel [132]	
Ca	26.70	36.38	36.44
P	12.47[a]	17.35	17.60
CO$_2$	3.48[b]	2.08	2.40
Na	0.731	0.699	0.70
K	0.055	0.031	0.045
Mg	0.436	0.21	0.21
Sr	0.035	—	—
N	4.92	0.064	0.058
Loss at 68°	—	2.12	1.58
Loss at 500°	—	2.8	2.5
F	0.07	—	—
Cl	0.08	—	—

[a] Expressed as PO_4^{3-}. [b] Expressed as CO_3^{2-}.

phase — something which probably has never been accomplished! Thus, the analytical results on dental enamel are of even greater interest than those of bone because the amount of organic matter present in enamel is comparatively insignificant. Analyses by LITTLE [132] for two samples of human enamel are shown in Table 10.1. The principal difference between the two samples was their surface areas: 18 and 28 m²/g, respectively. By way of commentary, it is noticeable that the first analysis shows greater amounts of loss (water?) and lesser amounts of both CO_2 and P, and thereby is consistent with the deduction that water (as H_3O^+ or as $H_4O_4^{4-}$) enters the structure of the inorganic phase of dental enamel. (Values reported by LITTLE for Ca are those obtained using a flame photometer. Other values for Ca were slightly lower if adjusted for Mg.)

LITTLE [132] observed a preferential leaching (removal) of carbon dioxide from dental enamel. From consideration of the surface areas involved, she concluded that the carbon dioxide was evenly distributed throughout the entire solid material, and that "lack of preferential removal of CO_2 from enamel can no longer be used to support a different mineral structure for enamel, bone or carbonate-apatites". Preferential leaching of particular ions — such as the removal of alkali from a feldspar — had been interpreted to imply that the carbon dioxide was adsorbed in some mysterious way on the surfaces of the very small crystallites. It is well known among geologists, however, that the basis of chemical weathering is the preferential dissolution of certain constituents in contradistinction to others [141].

In connection with such analyses, it becomes of interest to consider the composition of fossil dental enamel (dahllite), which probably contains even less organic matter than the fresh material reported by LITTLE [132]. The sample analyzed was a post-Wisconsin mastodon tooth found near Bluffton, Ohio [145], and the results are as follow:

CaO	MgO	Na_2O	CO_2	H_2O^+	H_2O^-	P_2O_5	Cl	F	Others
51.44	0.34	0.80	2.72	2.83	0.80	39.92	0.42	0.03	0.19

(The sum, corrected for (Cl + F), is 99.39; others are Al_2O_3 0.07, Fe_2O_3 0.03, K_2O 0.05, insoluble 0.04, SO_3 trace, and N_2O_5 trace.)

This analysis indicates twice the amount of water that is required to fill the two OH positions of hydroxyapatite!

This "excess" water has been considered by LITTLE & CASCIANI [133] who claim to recognize three different increments of chemically combined water in dental enamel, the last portion being removable only between 900° and 1300° C. The effects of such chemically combined water on physical properties — particularly the unit-cell dimension a have been considered under carbonate apatites, and the relation of water content to dental caries will receive further consideration.

DIBDIN [47a] has obtained NMR absorption spectra of dental enamel at 4 and at 97 per cent relative humidities and suggests that there is some "strongly bound water" which is not removed by simply lowering the relative humidity. It would be most interesting to know whether any changes in the unit-cell dimensions had occurred during removal of the "normal water".

10.2. Teeth and Bones: Possible Precursors

It has been assumed by several investigators that another crystalline (or so-called amorphous) phase exists as a precursor for the carbonate hydroxyapatite of bone. This assumption has been based primarily on Ca/P ratios below 1.667 and the further assumption that a small amount of a precursor always remains present. The argument is not logically consistent, however, because the same persons claim to have produced synthetic "calcium-deficient apatites", which would necessarily have Ca/P ratios below 1.667. There is no consistency, furthermore, with respect to explanations of the nature of this supposed precursor. An early suggestion [26] was $Ca_4H(PO_4)_3$, but FRANCIS & WEBB [68] suggest that it is brushite, $CaHPO_4 \cdot 2\,H_2O$.

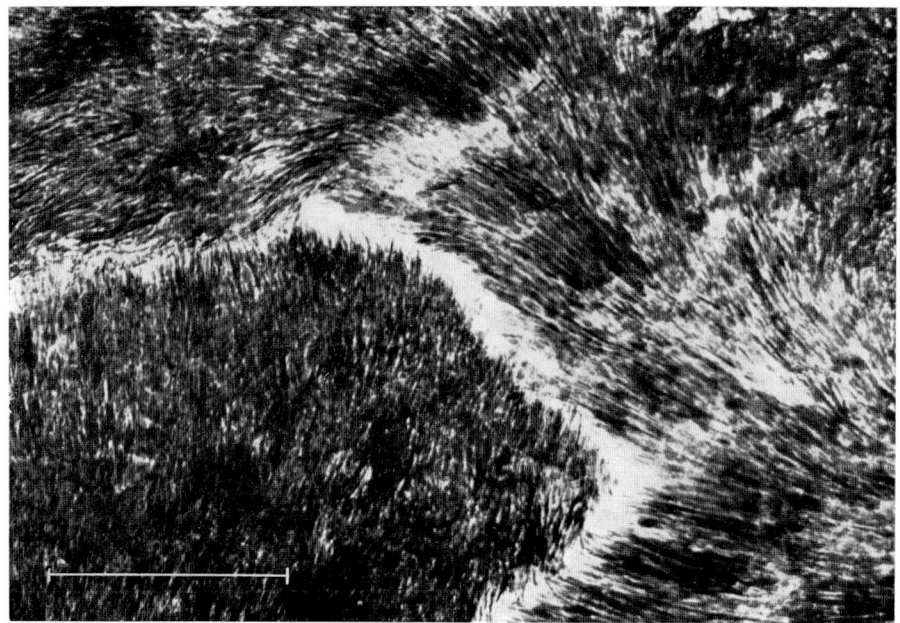

Fig. 14. Fetal dental enamel [human] showing early development of groups of crystallites in fluxional (subparallel) arrangement. There is no evidence of a so-called amorphous inorganic substance. The scale represents 1 μm. (McCONNELL & FOREMAN [161 a].) Copyright 1971 by the American Association for the Advancement of Science

While our own experience has shown that brushite may form as small, well-developed crystals in solutions approximating the inorganic composition of blood serum, our further experience indicates that they may redissolve and are not necessarily converted to apatite. It is precarious, however, to base conclusions on such inadequately characterized systems. One such example involved determination of the critical concentration of phosphate ions of the solution completely disregarding the fact that an organic phosphate had been employed as a buffer.

From the standpoint of pure conjecture the most incongruous is the "amorphous" substance which is supposed to comprise 40% or more of the inorganic portion of bone. Before discussing this fallacy, it is perhaps necessary to examine the meaning of this term, amorphous. To most informed persons it means that the substance is not only devoid of crystalline [external] morphology but that the internal structure is very poorly organized — not better organized than a commercial glass. It must be emphasized that the use of "amorphous" in connection with bone certainly carries with it the necessity that bone consists of two or more solid inorganic phases, because one of these phases is undeniably apatite.

There is no direct evidence from electron microscopy and electron diffraction of the presence of a second nonapatitic phase (crystalline or amorphous) in *any* of the biological precipitates that have been examined in our laboratory, and it can be said with certainty that no such quantity as 40% of an "amorphous" substance is present! Were such a quantity of an "amorphous" inorganic solid to be present in immature bone it should also occur in fetal dental enamel (see Fig. 14) for the same reasons.

How, then, were the erroneous conclusions obtained? The reasoning — or lack of it — is approximately as follows: If it is assumed that a crystalline aggregate is composed of individual single crystals which are of sufficiently large size to produce ideal resolution of the diffraction maxima, and if these crystals represent a well-known structural type (space group $P\,6_3/m$, in this case) and if there are no defects produced by compositional substitutions, then it follows that the powder diffraction pattern will be characterized by excellent resolution. (Bone produces a pattern which falls somewhat short of this ideal with respect to resolution, of course.) A standard can then be set up with which the observed deficiencies in resolution can be compared and thereby obtain semiquantitative values which supposedly represent the "degree of crystallinity". By subtracting this estimated number from 100% one can obtain a value for the "amorphous" material present — at least this is the crux of the argument.

One must consider, however, the three assumptions on which the argument is based. First, the size of the individual crystallites of bone is somewhat smaller than would be considered ideal. My colleague, Dr. FOREMAN, has observed with the electron microscope individual crystallites which show reasonable consistent shapes but are of the dimensional range of two to four times the a periodicity and ten to twelve times the c periodicity. Although diffraction effects from x-rays did not show good resolution, the improvement was significant with respect to the diffraction produced by the electron beam, and the orientation in subparallel arrangement of the crystallites was obvious from the degeneration of the "rings" into "segmental arcs" on the diffraction photograph.

Second, there are experimental data, as well as theoretical reasons, that indicate why the structure of the apatitic phase of bone should not be hexagonal. So the assumption of the space group is erroneous.

Third, the mere fact that *all* vertebrate bones contain a few per cent of carbon dioxide on analysis, and that the Ca/P ratio does not correspond with the ideal value (1.667) must be accepted as proof that ideal resolution of the diffraction maxima is not to be expected. To be sure, the proponents of "amorphous" bone were unwilling to concede that the CO_3 groups were present in the structure of the apatitic phase at the time this "amorphous" argument was first introduced. Unfortunately, this concept has been espoused by numerous persons who seem unaware of the limitations of the other methods — such as IR absorption — that supposedly supplement their x-ray diffraction conclusions.

On the other hand, the resolution of the diffraction pattern of bone is not as poor as has been implied by some. It is not worse than that of dentin (Fig. 15) provided advantage of the cumulative effects of the photographic method are employed and provided several possible refinements in technique are used for specimen preparation. The upper diagram (Fig. 15) represents our diffraction pattern of a synthetic carbonate hydroxyapatite prepared by KLEMENT [114]; it was prepared by heating appropriate constituents in a fused-quartz vessel to 400° C in the presence of carbon dioxide. The samples of *Lingula* and human dentin were *not* heated.

Fig. 15. Comparison of powder diffraction patterns of (a) synthetic carbonate apatite, (b) marginal portion of the shell of *Lingula*, and (c) human dentin. (McCONNELL [150])

In general, however, bone — and dentin — produce diffraction patterns inferior to those of dental enamel. The patterns of fossil dental enamel of a mastodon tooth (dahllite) [153] and a synthetic hydroxyapatite are compared (Fig. 16). In both specimens the sizes of the individual crystallites were sufficient to permit instrumental methods, rather than preparation of photographs. The dahllite specimen exhibits somewhat greater diffuse scattering, and this effect is probably attributable to a combination of characteristics, i.e., greater departures from ideal crystalline symmetry because of CO_3 groups within the structure, as well as a greater range in size of the individual crystallites.

While electron diffraction patterns (Fig. 17) are superior to those produced by x-ray diffraction, the possibility of having altered the nature of the original substance of bone cannot be ignored. The very low pressure of the chamber and the heat produced by the electron beam are conditions which cannot be dismissed when working with hydrated substances which also contain carbon dioxide. In fact, my colleague, Dr. FOREMAN, has observed a decrease in a for dental enamel amounting to more than 0.05 Å on comparison of the same material by x-ray and electron diffraction.

Returning to the question of the existence of a precursor, without specific reference to what it might be, the topic remains complex because of the ancillary theory which accompanies these premises concerning the nature of the precursor. In an attempt to demonstrate this function for brushite, for example, FRANCIS & WEBB [68] became absorbed with a possible epitaxic relation between brushite and hydroxyapatite.

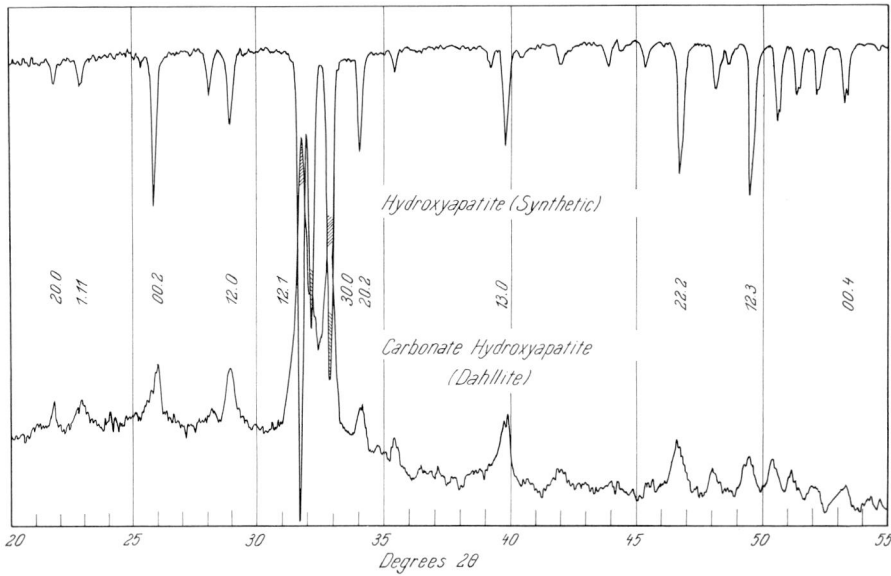

Fig. 16. Comparison of powder diffraction data obtained with the Philips apparatus using filtered copper radiation. The $(hk \cdot l)$ indices are indicated for some of the more intense interference maxima from (20.0) through (00.4). (MCCONNELL [153])

There should be no need for guesswork in connection with presence or absence of a significant amount of brushite in a sample of bone. The diffraction maxima of brushite (at spacings of 7.59—7.58 and at 4.24 angstroms) are of high intensity and are not closely adjacent to apatite maxima. Without demonstrating that vestiges of brushite are present, their conclusion becomes meaningless: "Hydrated calcium monohydrogen phosphate is proposed on the basis of rate and composition of calcium phosphate formed and on crystallographic data to be a necessary seed for growth of hydroxyapatite in bone and teeth at physiological pH." The "crystallographic data" are, in fact, conjectures concerning a possible epitaxic relation. It has been admitted that a precipitation of brushite may occur in solutions that approximate the concentrations of *inorganic* ions in blood serum, and surely the possibility of other crystalline phases admixed with an apatitic phase is not denied, but the salient question is: What has this to do with bone?

The simple fact of the matter is that nobody has been able to demonstrate the presence of a second crystalline phase for bone samples that have not been treated in some drastic way, such as boiling in alkaline solutions or in ethylenediamine. (We have already relegated the "amorphous" phase to the realm of the mystics.) Surely account must be taken of the fact that a crystalline substance (carbonate hydroxyapatite, containing a range of water contents) which forms under a comparatively narrow range of conditions with respect to temperature, pH, and concentrations, might be altered crystallochemically if these conditions were drastically changed.

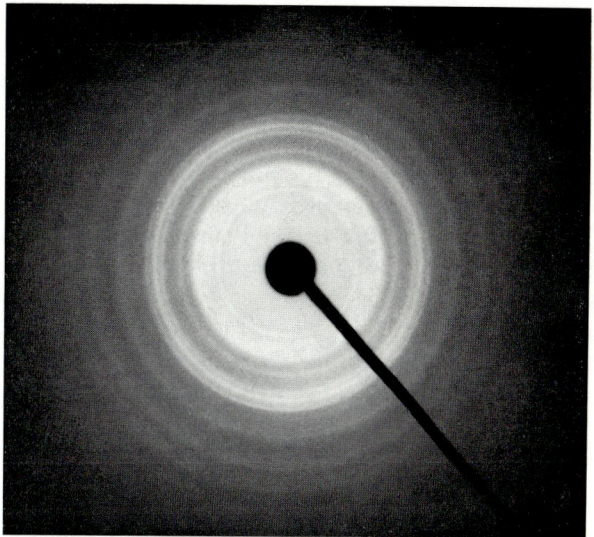

Fig. 17. Electron diffraction pattern of nondeproteinized bone, all maxima of which are attributable to dahllite (carbonate hydroxyapatite). The relative intensities have been altered by photomanipulation in order to enhance the weaker maxima at larger angles. (McConnell & Foreman [161 a].) Copyright 1971 by the American Association for the Advancement of Science

There is a vast literature on all aspects — including political — of the use of fluorides for reduction of caries susceptibility, particularly among adolescents. It is probable that fluorine enters the structure of the dental enamel, replacing the OH groups of the carbonate hydroxyapatite, and thereby contributes to a more stable structure, but the amount of fluorine is quite inadequate to account for the differences in the unit-cell dimensions [161]. *In vivo* studies suggested such a mechanism applies to incorporation of F in bone [13]. Aspects of fluorosis have been discussed by Largent [121].

Investigations of the "surface chemistry" of bone mineral supposedly have led to interpretations involving the crystal chemistry. Only one fundamental question arises: How can one determine the structure of a solid phase by dissolving it, heating it, measuring its surface area, or attempting to bring it into equilibrium with a solution of supposedly related composition? Polymorphism obviously precludes the determination of structure from even the most accurate knowledge of the composition, and this principle is elegantly demonstrated by the several crystalline forms in which silicon dioxide is known to occur, for example. Thus, while the use of isotopes — of Ca, P or even C or H — might reveal *in vitro* something about the physiological processes involved in bone replacement and regeneration, virtually nothing concerning the nature of the crystal chemistry of the carbonate hydroxyapatite (dahllite) has resulted.

Two decades ago, another substance called tricalcium phosphate hydrate (or alpha tricalcium phosphate) was widely discussed in the literature on bone. It was supposed to have a crystal structure very similar, if not identical, to hydroxyapatite. One of the principal lines of evidence for the occurrence of this phase in bone consisted of similarities in the x-ray diffraction patterns after thermal treatment. It is unnecessary to present contrary arguments concerning the existence of this mythical substance — in view of its subsequent demise — except to state that the arguments for its existence were quite uncomprehensible in terms of modern physical-chemical concepts in the first place. Nevertheless, in view of the tendency toward resurrection of completely disproven theories on the crystallo-chemical composition of bone, it surely should be mentioned. In this same connection, it should be noted that there has been a revival of the "double-salt hypothesis" that impaired some of HAUSEN's interpretations [90] prior to a knowledge of the structure of apatite. That HAUSEN's errors should be recurring about forty years later, merely indicates how difficult it is for those who are not qualified in crystal chemistry to make meaningful contributions.

10.3. Teeth and Bones: Mineralization

While the histological relationships of the living cells, osteoblasts, are discussed in considerable detail in many textbooks, comparatively little is known of the cytochemistry involved. As a metabolic function these cells presumably contribute some process which causes displacement of the equilibrium in such a manner that apposition of inorganic substance takes place. It has been implied on many occasions that these cells might contribute an enzyme which functions as a catalyst, and the enzyme most widely discussed in this connection is alkaline phosphatase.

However, it is interesting to note that another common enzyme, carbonic anhydrase, was found to play a primary rôle for *in vitro* systems. McCONNELL, FRAJOLA & DEAMER [162] were able to obtain precipitates of carbonate hydroxyapatite quite readily when a few parts per million of the crystalline enzyme was added, whereas no such precipitate was obtained during comparable time in the absence of the enzyme. Furthermore, when an inhibitor of the enzymatic activity — such as sulfanilamide — was added in addition to the enzyme, no precipitate was obtained during comparable time.

While these experiments do not demonstrate that the mineralization of bone is accomplished *in vivo* solely in the presence of carbonic anhydrase, there are other lines of evidence which imply that this enzyme may be related to precipitations of biologic apatite: such as the occurrence of carbonic anhydrase in saliva

[49] and in various situations where calculus or other phosphatic mineralization processes occur, including the mantle tissue of shellfish [85].

Indeed, the topic of mineralization (or calcification) [221] is most intricate. However, using a complex synthetic medium, *Bacterionema matruchotii* will produce intracellular depositions comparable to that of the inorganic component of bone [56]. The medium contained nine vitamins; pimelic and thioctic acids; casein hydrolysate; and adenine, guanine, thymine, uracil and xanthine; in addition to inorganic components and a buffer. While these experiments with bacteria of various species [196] may be directly related to formation of oral calculus, it is difficult to believe that they will lead to general clues concerning bone mineralization in the immediate future. At least some bacteria — notably Neisseria — indicate carbonic anhydrase activity [226]. Normal bone development occurs for laboratory animals bred under germ-free conditions, however.

Inhibitors, in contrast to catalysts, apparently exist also, and Mg ions seem to retard or impair crystallization for *in vitro* systems [216], as do also pyrophosphate and polyphosphates [63] in some *in vivo* situations [199].

If, however, it can be assumed that the crystallochemical properties of dahllite have been established in a general way in accordance with the principles outlined under carbonate apatites, it is possible to extend this information to at least one pathological condition, i.e., dental caries. It has been found [109, 161] that caries-susceptible and caries-resistant dental enamels yield measurably different dimensions for their a periodicities, as ascertained by x-ray diffraction methods, and the suceptible group exhibit larger a dimensions. What does this indicate concerning the chemical differences?

Again, it becomes necessary to re-examine current knowledge on the crystallochemical characteristics of carbonate apatites, and it becomes evident that the only constituent that could produce such a range of a periodicities is the most elusive from the viewpoint of chemical analysis, i.e., chemically combined water. To be sure, the effects of fluorine in control of dental caries are not excluded by this discovery [161], but the observed differences cannot be accounted for in terms of the amounts of fluorine *per se*. Indeed, it has been suggested as early as 1952 [140] that some of the diverse properties of dahllite were related to variations in the chemically combined water.

10.4. Teeth and Bones: Trace Elements

McLean & Budy [168] have devoted major portions of a book to use of isotopes in experimentation on bone *in vivo* and *in vitro*. Sr-90 is potentially hazardous insofar as it is a bone-seeking element that has a relatively long half-life. Fortunately there is discrimination against Sr in favor of Ca, both *in vivo* and *in vitro* [181]. In addition to small amounts of sodium, magnesium and potassium, more than 20 other elements are found as traces [168]; among these zinc has attracted attention because it is believed to both promote and inhibit mineralization. Vincent [229] believes that zinc does not enter the structure of dahllite but is present as "part of a metallo-enzyme in the mineralizing mechanism". One notes that both alkaline phosphatase and carbonic anhydrase are zinc-containing enzymes [222].

While some trace elements — particularly Fe, Cu, Mn, Zn, I and Co and possibly Mo, F, Ba and Sr — are believed to be essential to metabolic processes of higher forms of animals, little is known with respect to their rôles in bone mineralization or their abilities to enter the dahllite structure. In connection with dental enamel, fluorine is an exception, of course, as has been previously indicated. There can be little doubt that part, if not all, of the Na, Mg and K could enter the apatite structure. Despite some offhand assumptions concerning the differences in ionic radii, such weight percentages as 0.7 for Na and 0.06 for K have been greatly exceeded in synthetic preparations of SIMPSON [205]. The substitution of Mg for Ca in magnesian calcites involves amounts as great as 7.8 weight per cent of MgO for some calcareous algae [75], so it reasonably can be assumed that 0.4% of Mg could occur in the apatite structure. Although such amounts of Mg do not seem to be common, 0.53% of MgO is reported for an analysis of francolite [228].

Table 10.2. *Elemental Analyses of Human Teeth*

	Analytical Results		Normalized	
	Dentin[1]	Enamel[1]	Dentin	Enamel
Ca	26.2%	37.0%	100%	100%
Mg	0.87%	0.28%	3.3	0.76
Na	0.55%	0.70%	2.1	1.9
Sr	94 ppm	111 ppm	359	300
Zn	173 ppm	263 ppm	660	711
Ba	129 ppm	125 ppm	492	338
Fe	93 ppm	118 ppm	355	319
Al	69 ppm	86 ppm	263	232
Ag	2 ppm	0.6 ppm	7.6	1.6
Cr	2 ppm	1 ppm	7.6	2.7
Co	1 ppm	0.1 ppm	3.8	0.3
Sb	0.7 ppm	1 ppm	2.7	2.7
Mn	0.6 ppm	0.6 ppm	2.3	1.6
Au	0.07 ppm	0.1 ppm	0.27	0.27
Cl	0.035%	0.32%	0.13	0.86
Br	114 ppm	34 ppm	435	92

[1] In general, the values given here represent reduction by one or more significant figure from those given by the authors [195 b].

Using neutron activation, RETIEF et al. [195b] have analyzed for fourteen metallic elements of "normal" human teeth; their results are summarized in Table 10.2. While the analytical methods seem acceptable, these isolated data were confined to ten (or fewer) determinations for each element. No information is given concerning the previous owners of these teeth with respect to sex, age, place of residence, etc. The differences between values for dentin vs. enamel become more evident when related through normalization to the amount of calcium. In the case of Mg, for example, its greater abundance in dentin, by a factor four, might imply that some of the Mg of dentin was present as either an organic compound or an evaporated liquid, rather than being contained within the structure of dahllite. A similar situation seems probable for Br, but quite the opposite is true for Cl. The contents of Na and Mg for enamel are essentially in agreement with those in Table 10.1.

Speculation concerning electric currents and their rôle in bone apposition or resorption quickly leads to the question: Are the individual inorganic crystallites piezoelectric? And while the answer must surely be negative if these crystallites have the structure of *Fap*, it is recognized that *Hap* cannot belong to the same space group and might have a polar axis. BASSETT [11] has reviewed some of the concepts involved and has suggested that an electromotive force could arise from junction potentials at crystalline-collagen interfaces. There can be little doubt that electrochemical processes are active in most physiological situations, but the experimental limitations make *in vivo* measurements extremely difficult to obtain and interpret.

10.5. Other Biologic Precipitates

Among invertebrates, inorganic phosphatic depositions are restricted to a few phyla insofar as current knowledge is concerned. LOWENSTAM [135] indicated phosphatic skeletal tissues for certain mollusks and arthropods, and CLARKE [36] gave analyses of crabs and shrimps which show a range from 15—50% when calculated as tricalcium phosphate. One cannot assume, however, that a carbonate apatite is necessarily present because several other phosphate biominerals cannot be excluded as possibilities. However, the inarticulate brachiopod genus, *Lingula*, has been investigated; it has a shell composed of francolite [150] and thereby differs from vertebrate phyla in its ability to accumulate fluorine.

A widely distributed class of small fossils called conodonts should be mentioned because of their apatite composition [52, 88]. They are extensively used for stratigraphic correlation by paleontologists, despite the dearth of knowledge concerning the nature of the organism that produced these denticular fossils. The apatitic composition *in vivo* is surmised, of course.

The scales of many, if not all, bony fish are apatite insofar as the inorganic component is concerned, as are the incipient crystallites to be found in the hyaline cartilage of the spinal column of the leopard shark [150]. The nature of the scales of the gar-pike has been investigated by CARLSTRÖM; he found the "ganoin" to be composed of a substance "typical for carbonate-containing biological apatites", as reported by ØRVIG [184, p. 85—85].

Pathological depositions of vertebrates frequently are apatitic in part if not essentially. LONSDALE [134], LAGERGREN [120] and PRIEN & FRONDEL [193] have summarized the work on urinary "stones" most thoroughly, and merely one additional precautionary comment probably should be repeated here: Liberation of gas from a renal stone does not indicate the presence of a carbonate phase because most, if not all, apatitic stones are carbonate-apatite varieties. On the other hand, it has been the writer's experience that the carbon dioxide content of apatitic phases has been overlooked by use of inadequate testing methods.

Induced cardiovascular mineralization in cattle [31] and induced corneal calcification in rabbits [59] are apatitic in character, as are at least some of the inorganic depositions within lungs associated with histoplasmosis, for example.

Although dental calculus is often essentially apatite, other phases (brushite, whitlockite, octacalcium phosphate and possibly monetite) have been reported. The possibility of contamination of a calculus sample by a dentifrice which contains a calcium phosphate must be considered, however, if one wishes to make deductions concerning the physiological chemistry of the oral cavity.

While the mineralization (or calcification) of osteoid tissue and remodeling (regeneration) and resorption of bone belong in the realm of physiological chemists [169], and consequently are not proper topics for this work, it should be stated most emphatically that it will not be feasible to interrelate the dynamics of bone formation to the biochemistry of organisms as long as bioscientists continue to equate the bone mineral with $Ca_{10}(PO_4)_6(OH)_2$. The inorganic phase of bone should *not* be called hydroxyapatite; it is a dahllite — provided a mineral name is appropriate for something of organic origin — or carbonate hydroxyapatite (*carHap*) if one is attempting to designate both the composition and structure.

11. Critique

The reader may have noticed in several instances what may have seemed to represent a pessimistic attitude toward the future of research on some of these topics. On occasion it might be suggested that the author should have used a subtitle such as: What is *not* known about apatite. Facetiously, the rejoinder to such an appraisal would have to be: One could not write a book on what is not known on this subject because many volumes would be required in order to accomplish such a task. However, in order to sustain an interest in research on any particular topic for more than three decades, one must surely have an optimistic viewpoint, despite the fact that the road he travels may be a most lonesome one.

The crystal chemistry and mineralogy of apatite necessarily could not progress at a more rapid pace than the development of a general body of theory in these two areas of knowledge. While it is true that some proposals were first advocated on the basis of data acquired through study of apatite, the situation was most discomforting during intervals when apatite appeared to be the principal, if not sole, example of a hypothetical crystallochemical postulation, as was the situation for a number of details during several years.

Most of the theoretical discussions in preceding chapters are based upon so-called classical experimental methods which now include diffraction techniques — certainly for the purpose of obtaining dimensions and ideal structures. When the structures show significant departures from the ideal, as is apparently true for carbonate apatites, the diffraction method loses much of its capability. Being more specific on this point: there are surely limitations to the extent of structural information that can be attained by study of a triclinic crystal if one begins with the erroneous premise that the crystal is hexagonal. This statement is not intended to imply any skepticism or disparagement with respect to logical analogies; they are extremely useful as long as they are general, at first, and become specific only after thorough justification.

Although the diffraction methods have proved of great service for nearly half a century and will retain a usefulness for determining dimensions, it is questionable whether much greater insight on structure of apatites can be attained by diffraction methods until somebody is successful in producing a single crystal of *carHap* or *carFap* of discrete dimensions. (In this connection the author is somewhat pessimistic, insofar as he has the intuitive feeling that such tiny crystals may have a most pronounced tendency to form twins.)

One might wonder why the author has given so little credence to data obtained by infrared absorption spectroscopy, and the answer to this question is to be found among the lack of reliability of most of the interpretations based on such data, coupled with their incompatibility with data obtained by other methods that are capable of more straightforward interpretation. It is true, of course, that FISHER & McCONNELL [62] considered the IR spectra as a basis for their conclusion that aluminum substituted for both calcium and phosphorus in the product obtained by heating morinite. However, far better justification for their conclusion originated from the chemical stoichiometry in conjunction with optical and diffraction data, so the IR data were merely confirmatory and in no way represented a primary basis for their conclusion.

What useful information the IR methods may ultimately yield remains to be demonstrated, but the use of polarized radiation may produce some details concerning *carFap* and/or *carHap* that are consistent with generalizations obtained by other methods. Up to the present, most of the interpretations have been highly speculative because of lack of knowledge of the limitations of such methods. For example, some of the earliest interpretations were so rash as to assume that one could ascertain that the CO_3 bands of *carFap* were produced by contamination with calcite merely because of superficial resemblance between the absorption data. (One would, of course, expect a CO_3 group to have certain characteristic frequencies which would be more or less independent of its environment, and the most cursory measurements could easily fail to discover frequency modifications produced by environmental factors.) Indeed, the inconsistency between interpretations obtained by infrared data and data obtained by use of other frequency ranges recently has been explained [130, 113]. It has been found essential to consider whether one is dealing with a lone wolf or some particular member of a pack of wolves, if one will permit allegory.

Nuclear magnetic resonance data have been obtained for *Fap* [90a] and paramagnetic centers have been found as an interpretation of electron spin resonance spectra [202], but whether such methods will permit major contributions in the future is uncertain. It is claimed [72b] that electron capture (or loss) produced centers CO_3^- and CO_3^{3-} for the carbonate replacing PO_4 of several samples of *carFap*, and ESR measurements have led to the conclusion that CrO_4^{3-} occurs in synthetic Cl-*ap*, according to BANKS et al. [10].

Such highly sophisticated methods may have value in confirming hypotheses that have already attained theoretical status for other reasons, but such interpretations remain highly questionable when they lead to conclusions contrary to more direct types of evidence.

It should be recognized that the demonstration of the pentavalent character of Cr required accurate knowledge of the space-group symmetry of the Cl-*ap* as $P2_1/b$. Use of such spectral methods to compare crystals of uncertain symmetry with a so-called amorphous [or glassy] substance is absurd.

Neutron diffraction is a most appropriate method for locating deuterons in highly symmetrical situations — provided they are sufficiently abundant! But one must surely avoid looking for a needle in a haystack, particularly if it is a large haystack, so the author ventures the opinion that little will come from application of neutron diffraction beyond the more precise location of the protons of ideal *Hap* [with nearly perfect stoichiometry] after somebody is successful in synthesizing $Ca_{10}(PO_4)_6(OD)_2$. One knows in advance that such a structure cannot have the true symmetry of $P6_3/m$ because OD cannot be centrosymmetric, so that any basis for locating a deuteron (in lieu of a proton) will require a realistic refinement of the structure of *Hap* as a starting point. Present knowledge does not even indicate whether *Hap* has a polar axis through precise examination by piezoelectric or pyroelectric methods. Indeed, the question is foremost whether an ideal crystal of *Hap*, of usable size, has been produced.

Moreno et al. [178a] prepared *Hap* and heated two samples of their product to 1000° C in air and steam atmospheres. They stated that both samples dissolved stoichiometrically, but produced activity products that differed by about 10^3. Their calculations include "allowance ... for the presence of the ion pairs $[CaHPO_4]°$ and $[CaH_2PO_4]^+$ ". They concluded that the steam-heated sample had "either a more uniform environment for the OH^- group in this sample or actually a higher content of OH^- in the crystals". It is significant that the mean refractive index of their *Hap* was merely 1.636, as compared with a mean of 1.649 [160]. Their work is noteworthy to the extent that it raises more questions than it answers.

The foregoing comments on physical methods are far from comprehensive, and it is with trepidation that the author now attempts to discuss chemical methods of analysis — some of which are so simple that an inexperienced technician can be taught to achieve reasonable precision and accuracy within a few days, but others of which are so complex as to result in little precision and consequently less accuracy. An example of the first sort is the determination of CO_2 on samples weighing about a gram through absorption of the liberated gas on Ascarite, or the use of smaller samples and a Warburg apparatus. With standardization such methods can produce both precision and accuracy.

Since the conclusion of the writer's formal instruction in analytical chemistry, many sophisticated methods have come into common use, including numerous physical methods, such as measurement of intensity of x-ray spectra. Electron probe analysis is surely appropriate for some situations and is capable of good accuracy for certain purposes. However, what is often absent in analytical reports is an indication that standardization has been accomplished in connection with classical methods and that the newer instrumental techniques are applicable in the range of compositions under consideration.

Most dismaying to the author is the lack of understanding of the significance of sampling methods and their relation to accuracy. Entirely too many inexperienced analysts do not even seem to recognize that an aliquot factor is also applicable to the error of determination. For example, methods intended for determination of phosphorus in urine have been used for determination of phosphorus in apatite by diluting a solution containing dissolved apatite to the approximate concentration range of urine. The dilution required is about one hundredth but the same absolute error for a determination (0.10 ± 0.01 or 10%) is applicable to the error of P for apatite. Thus, there is no need to calculate Ca/P ratios to three (or more) significant figures when the second figure is surely in doubt as a consequence of the method.

Another matter which seems to confuse some modern scientists is the difference between precision and accuracy. The former should be understood to mean some function of reproducibility, whereas the latter means the approach to the correct or true value. Inasmuch as there is no perfect method for obtaining the correct value, comparisons must be based on standard samples for which "true" values have been assumed. Establishment of such standards obviously requires cooperation of numerous individuals and statistical analysis of their results. Although such surveys are feasible for rock samples, including phosphorites, the available supply of a single mineral phase precludes wide application of such comparisons of methods in use in different laboratories in order to evaluate precision and accuracy. A few theoretical standards have been assumed, such as optical-quality calcite (CaO and CO_2) and quartz (SiO_2). These standards, particularly when subjected to emission spectroscopic examination may be quite acceptable, but most experienced analysts know that one does not depend upon the label of ordinary reagent chemicals except for more casual purposes. For example, the reagent grade "tricalcium phosphate" does not have a Ca/P ratio 1.5 and ordinarily will produce appreciable effervescence when placed in 3 N HCl because of its significant carbon dioxide content. Furthermore, most of the intense diffraction lines which it produces are attributable to apatite.

One of the most severe limitations encountered in the analysis of an apatite is the determination of water, which is very tenaciously retained. There is, of course, a relation between time and temperature, but it is highly questionable whether virtually all water can be removed below 1100° C during any reasonable amount of time. Even when using recommended fluxes [41] in conjunction with PENFIELD's method, recovery of all the water is uncertain, and it is advisable to perform an ignition to 1200° or upwards as a check on the total volatile constituents. If the loss on ignition significantly exceeds the sum of $CO_2 + H_2O + 1/2$ ($Si F_4$), suspicion should be directed toward the water determination.

As mentioned previously, the oxyapatite (voelckerite) problem is related to the accurate determination of structural water [164], rather than argumentation about the valence of Pb, for example.

Thus, one of the clearly essential needs is a chemical method for determination of water in mineral and biologic apatites. While the IR absorption band at about $3250-3650$ cm^{-1} (2.74—3.1 nm) is quite sensitive as a qualitative method, it does not seem to have good potentialities for quantitative accuracy, insofar as can be predicted.

ARMSTRONG and his associates [225a] have recently used the KARL FISCHER titration method in an attempt to ascertain the water contents of so-called anorganic bone [deproteinized by refluxing with ethylenediamine] and human dental enamel. Prior to their determinations of water, however, they dried the samples to remove the "molecular water bound to apatite" by heating to 161—164° C in vacuum for 25—27 hours. When dental enamel was brought to equilibrium with the laboratory air, however, the amount of water was greater by a factor of 1.38. The results on anorganic bone yield a much larger factor (when compared with 1.38) namely, 2.19.

In relation to the mineral specimens, such as francolite which contains more than the theoretical amount of fluorine, and contains additional water that surely is *not* "molecular water bound to apatite", the use of such drying procedure surely requires greater justification. Thus, although a KARL FISCHER titration may be an appropriate method for determination of water in apatites, the question will not be resolved with respect to bone until one is capable of resolving the enigma of the ancient expression "bone dry". The probable alteration of the composition of the inorganic portion of bone by ethylenediamine has been mentioned.

A general commentary on this work [225a] and on similar works is: it seems more appropriate to fit the hypothesis to the data rather than make the data fit the hypothesis — particularly when the hypothesis is likely to be erroneous. Thus, such expressions as, "Bone mineral has been described as microcrystals of hydroxyapatite with surface-located ions not involved in the crystal lattice structure", should not be used as an initial premise. The "surface-located ions" definitely produce changes in the unit-cell periodicities, and thus their lack of involvement with "the crystal lattice structure" becomes impossible to explain on the basis of any crystallochemical theory with which the author is familiar. Furthermore, as previously explained, bone should *not* be referred to as hydroxyapatite.

Appropriate at this juncture might be some comments on semantics. Bioscientists have been quite content — if not eager — to use such terms as mineralization, apatite, crystal chemistry, unit cell, lattice, solubility product, equilibrium, etc., but in adopting such terminology they have been most careless with respect to appropriate connotations and denotations. While the author has no objection to the use of "mineralization" for the formation of an inorganic substance in tissues or organs, for example, it should be recognized that this term has a different connotation in connection with petrologic paragenesis, where its use considerably antedates its use in connection with hard tissues. Thus, the key word, mineralization, acquires two entirely different meanings during a mechanized search for information.

The author does object, however, to (1) attempts to calculate a "solubility product" for an aqueous system that is not at equilibrium, (2) complete disregard of the definition of "phase", and (3) failure to consider the number of phases present in multicomponent systems. In addition he strongly objects to the use of "ion exchange" in a vague, general way to include any and all types of chemical reactions, and to the intriguing supposition that merely because a statement using chemical symbols can be written, such a reaction unquestionably does (or can) occur. For example, one can write an expression for the dissolution of albite in water:

$$2\,NaAlSi_3O_8 + H_2O \rightleftharpoons 2\,NaOH + Al_2O_3 + 6\,SiO_2.$$

Any question about the dissolution of albite in water is answered by the phenomenon of chemical weathering; the two substances on the left do react and with a velocity that can be measured within months or even weeks. What forms as a result of this dissolution is surely in doubt. Almost certainly Al_2O_3 and SiO_2 do not form as simple anhydrous oxides. Whether the equilibrium can be displaced to the left is highly questionable if given a time interval less than a million years at atmospheric temperatures. We conclude, therefore, that the above statement is nonsense, as are many statements regarding various phosphate reactions.

Indeed, the author has not made himself particularly popular among health scientists on previous occasions by calling their attention to unrealistic approaches involving absorption theories that deviate completely from a crystallochemical approach. The situation is reminiscent of attitudes which were quite popular for the explanation of the properties of clay minerals a few decades earlier. One simple fact persists in both situations: the substances are definitely crystalline, and despite their high surface areas because of the extremely small sizes of individual crystallites, fundamental explanations for their behaviors must be based upon sound crystallochemical theory. Dental enamel, for example, produces far better powder diffraction patterns than does the clay mineral montmorillonite for which it has been feasible to explain most of the mineral's exceptional characteristics in terms of crystallochemical principles, including its high cation exchange capacity, its behavior as a mild oxidizing agent in conjunction with interaction with certain dyes, its ability to interact with organic molecules, its variation of the c axial dimension, and the disordering of the structural sheets and consequent loss of hkl reflections on the diffraction pattern.

The problems involving the inorganic component of hard tissues are dwarfed, in some ways, by those which confronted the clay mineralogists, so there is every reason to believe that far greater knowledge will surely become available on the composition of bone. The earlier clay mineralogists — more appropriately called soil scientists — had an "amorphous" component that was used to fill in a niche between empiricism and theory, but with the disappearance of the niche, the "amorphous component" designation seems to have lost its place to a more helpful concept defining specific disordering of the ideal structures [22].

The real challenge remaining is that of interrelating the precipitation of dahllite to the vital processes of some invertebrates as well as all vertebrates, whether normal or pathological in nature. The problem is not unrelated to direct precipitation of phosphorite from sea water [154, 180a] or to precipitation of apatite by bacteria [56, 196]. Inasmuch as the carbonate ion is present in all such solid substances, no progress will occur on this topic as long as this component — of both the solution and the solid — is disregarded [49a].

Appendix

Data on Apatites and Related Substances

Mineral name: *brushite*
Compositional-structural name: dicalcium phosphate dihydrate
Formula: $CaHPO_4 \cdot 2H_2O$
Optical properties: biaxial $+$; $\gamma = 1.553$, $\beta = 1.546$, $\alpha = 1.539$, $2V = 69$ to $87°$.
Density: 2.30—2.31.
Crystallographic data: monoclinic, $A2$; (synthetic) $a = 5.81$, $b = 15.18$, $c = 6.24$, $\beta = 116.4°$, $Z = 4$.
Prominent diffraction spacings, intensities to the base (10): (10) 7.57 and 4.24, (8—7) 3.05, (5) 2.93 and 2.62, (3) 2.603 A.
Occurrences: Bioprecipitates, including uroliths, and both insular and continental phosphate deposits.
Remarks: Brushite is structurally similar to gypsum for which the optic angle shows pronounced change within the temperature range $0°$ to $90°$ C, but unlike gypsum its solubility does not increase at lower temperatures. It has been suggested that this substance is not stable above $25°$ C, except in contact with a solution.

Mineral name: *chlorapatite*
Compositional-structural name: chlorapatite
Formula: $Ca_{10}(PO_4)_6Cl_2$
Optical properties: biaxial or uniaxial neg.; $\gamma = \omega = 1.668$, $\alpha = \varepsilon = 1.665$.
Density: 3.17.
Crystallographic data: $a = 9.64$, $c = 6.78$ (if hexagonal).
Prominent diffraction spacings, intensities to the base (10): (synthetic) (10) 2.78, (6—5) 2.86, (2—1) 3.38, 2.32, 1.964, 1.843, 1.824 and 1.835 A.
Occurrences: Veins associated with basic rocks, iron ores of complex origins, and silicified marbles.
Remarks: While many apatites show small quantities of Cl, a close approach to the end member ($Ca_{10}(PO_4)_6Cl_2$) seems to be rare. Monoclinic Cl-*ap* has been reported [97] with space group $P2_1/a$, $a = 19.210$, $b = 6.785$, $c = 9.605$ and $\beta = 120°$. This material had $2V \sim 10°$, which is consistent with biaxial properties reported for some synthetic products [192]. The refractive indices are for the natural material [97].

Mineral name: *dahllite*
Compositional-structural name: carbonate hydroxyapatite
Formula: $(Ca, X)_{10}(P, C)_6(O, OH)_{26}$. With $CO_2 > 1\%$ and $F < 1\%$ by weight.
Optical properties: mean $n \leq 1.610$ to ≥ 1.520, usually $1.59 - 1.55$.
Density: <3.10 but >2.96.
Crystallographic data: pseudohexagonal; $a < 9.45$, $c > 6.90$, depending on CO_2 and water content.
Prominent diffraction spacings, intensities to the base (10): (>10) a doublet at 2.81, (8) 2.73, (3) 3.44, 3.10, 2.27 and 1.95 A.
Occurrences: Biologic hard tissues, insular rock phosphates and as a secondary (weathering?) deposition in cavities of various rocks.
Remarks: The physical properties show a considerable range with respect to changes in both CO_2 and structural water. The range of n represents reported extremes [193], whereas the prominent diffraction lines are for fossil dental enamel [145].

Mineral name: *ellestadite* (including hydroxylellestadite)
Compositional-structural name: apatitic sulfate-silicate of calcium
Formula: $Ca_{10}(SiO_4)_3(SO_4)_3(OH, F, Cl)_2$
Optical properties: uniaxial neg.; $\omega = 1.654$, $\varepsilon = 1.650$.
Density: $3.02 - 3.08$.
Crystallographic data: hexagonal, $P6_3/m$; $a = 9.48$ (corrected for Cl), $c \sim 6.92$ A.
Prominent diffraction spacings, intensities to the base (10): (10) 2.84, (6) 2.74, (5—4) 2.80, 2.65 and 1.853, (4) 3.46 A.
Occurrences: Calcic-silicic metamorphic rocks.
Remarks: Synthetic $Ca_{10}(SiO_4)_3(SO_4)_3F_2$ has $a = 9.43$ and $c = 6.93$ [119]; the large dimension ($a = 9.48$) is for ellestadite containing a significant amount of structural water, but has been reduced to correct for Cl.

Mineral name: *fluorapatite*
Compositional-structural name: fluorapatite
Formula: $Ca_{10}(PO_4)_6F_2$
Optical properties: uniaxial neg.; $\omega = 1.632$, $\varepsilon = 1.628$.
Density: 3.199 (calculated).
Crystallographic data: hexagonal, $P6_3/m$; $a = 9.37$, $c = 6.88$.
Prominent diffraction spacings, intensities to the base (10): (10) 2.80, (7—6) 2.70, (3) 2.25, (3—2) 2.77, 3.06, 1.835 and 1.935 A (synthetic).
Occurrences: Igneous and metamorphic rocks, primary mineral of pegmatites, and to a limited extent as detrital grains in sediments.

Mineral name: *francolite*
Compositional-structural name: carbonate fluorapatite
Formula: $(Ca, X)_{10}(P, C)_6(O, F)_{26}$. With $F > 1\%$ and $CO_2 > 1\%$ by weight.
Optical properties: usually biaxial, mean $n \leq 1.630$, $\varDelta \geq 0.005$.
Density: <3.18, usually $3.15 - 3.13$.
Crystallographic data: pseudohexagonal; $a \leq 9.36$, $c < 6.89$, depending on CO_2 and water content.

Prominent diffraction spacings, intensities to the base (10): (10) 2.79, (7) 1.833, (6) 2.70, (4) 3.44, 2.77, 1.93, 1.79, 1.77, 1.74 and 1.72 A.

Occurrences: Principal mineral of rock phosphates, hydrothermal veins, and as a biomineral forming the shell of *Lingula*. Possibly a primary mineral of other brachiopods and conodonts.

Remarks: The physical properties depend on the contents of CO_2, F and structural water. For example, Δ may even exceed 0.016. The prominent diffraction lines are for the occurrence at Kutná Hora [95]. The optic angle is usually small, but maybe as large as 40° [45].

Mineral name: *hydroxyapatite*
Compositional-structural name: hydroxyapatite
Formula: $Ca_{10}(PO_4)_6(OH)_2$
Optical properties: uniaxial (?) neg.; γ or $\omega = 1.651$, α or $\varepsilon = 1.647$, $2V \sim 0°$ (synthetic).
Density: 3.156 (calculated).
Crystallographic data: hexagonal (?), possibly monoclinic with $\beta = 120°$; $a = 9.42$, $c = 6.88$.
Prominent diffraction spacings, intensities to the base (10): (10) 2.81, (8—7) 2.72, (7—6) 2.78, (5—4) 1.845, (4—3) 3.44 and 1.943 A.
Occurrences: Various metamorphic rocks, but probably always as an isomorphic variant containing F and/or Cl, and possibly CO_2.
Remarks: Isomorphic variants closely approaching $Ca_{10}(PO_4)_6(OH)_2$ are not known to occur naturally. All data given here are for synthetic preparations, and it is questionable whether a crystal of *Hap* has ever been obtained with adequate size to permit single-crystal measurements. (The latter statement stands in contrast to some claims based on casual experimental methods.)

Mineral name: *monetite*
Compositional-structural name: dicalcium phosphate (anhydrous)
Formula: $CaHPO_4$
Optical properties: biaxial neg.; $\gamma = 1.635$, $\beta = 1.613$, $\alpha = 1.586$, $\Delta = 0.052$, $2V = 84°$.
Density: 2.89.
Crystallographic data: triclinic, $P\bar{1}$; $a = 6.90$, $b = 6.65$, $c = 7.00$, $\alpha = 96.23°$, $\beta = 103.9°$, $\gamma = 88.97°$, $Z = 4$.
Prominent diffraction spacings, intensities to the base (10): (10) 2.958, (8—7) 3.35, (7) 3.37, (4—3) 2.937 and 2.721 A.
Occurrences: Insular phosphatic rocks associated with guano deposits.
Remarks: Monetite will re-hydrate to form brushite, presumably under atmospheric conditions of temperature and relative humidity.

Mineral name: *morinite*
Formula: $\sim Ca_4Na_2Al_4(PO_4)_4(F, OH)_7 \cdot 5H_2O$
Optical properties: biaxial neg.; $\gamma = 1.565$, $\beta = 1.563+$, $\alpha = 1.551$, $2V = 43°$.
Density: 2.96.

Crystallographic data: monoclinic, $P2_1/m$; $a = 9.46$, $b = 10.69$, $c = 5.45$, $\beta = 105.46°$.
Prominent diffraction spacings, intensities to the base (10): (10) 2.94, (8) 3.47 and 1.783, (7) 4.70 and 2.63, (6) 9.11 and 3.73 A.
Occurrences: Pegmatites, during intermediate stages of crystallization.
Remarks: When heated morinite recrystallizes to form Al-rich apatite and, under some conditions, a whitlockite phase.

Mineral name: *strontiapatite*
Compositional-structural name: strontiapatite
Formula: $(Sr, Ca, et\ al.)_{10}(PO_4)_6(F, OH)_2$ ($SrO = 46.06\%$)
Optical Properties: uniaxial neg.; $\omega = 1.651$. $\varepsilon = 1.637$, but for synthetic $Sr_{10}(PO_4)_6F_2$, $\omega = 1.621$, $\varepsilon = 1.619$ and $\Delta = 0.002$.
Density: 3.84 (synthetic 4.19).
Crystallographic data: hexagonal; $a = 9.66$, $c = 7.19$ kX. For synthetic $Sr_{10}(PO_4)_6F_2$ $a = 9.70$, $c = 7.26$ kX.
Prominent diffraction spacings, intensities to the base (10): (10) 2.89, (7) 3.17, 2.78, 2.005 and 1.909, (6) 1.467, (5) 2.32 and 2.147 kX.
Occurrences: High-strontium apatites are rare; they seem to be confined to pegmatites and high-temperature veins.
Remark: Synthetic $Sr_{10}(PO_4)_6F_2$ has very low birefringence, whereas that of the natural substance is unaccountably high. The diffraction lines given (in kX) are for the natural substance [248]; conversion to A involves multiplication by 1.00202.

Mineral name: *whitlockite*
Compositional-structural name: β tricalcium phosphate
Formula: $Ca_3(PO_4)_2$
Optical properties: uniaxial neg.; $\omega = 1.629 - 1.607$, $\varepsilon = 1.626 - 1.604$, $\Delta = 0.003$.
Density: 3.12−2.90.
Crystallographic data: hexagonal, $R\bar{3}m$; $a \sim 10.3$, $c \sim 36.9$, $Z = 21$ [212].
Prominent diffraction spacings, intensities to the base (10): (10) 2.880, (7−6) 2.607, (6−5) 3.21, (3−2) 3.45, (2) 5.21 and 2.757 A.
Occurrences: Pathologic precipitates, associated with dahllite in insular phosphate rocks and as a hydrothermal product in pegmatites.
Remarks: Whitlockite may contain small amounts of MgO and CO_2; these and other compositional variations (including water) presumably account for the range for density and for refractive indices [75a].

Mineral name: *wilkeite*
Compositional-structural name: isomorphic variant between apatite and ellestadite.
Formula: $Ca_{10}(P, Si, S)_6O_{24}(OH, F, Cl)_2$
Optical Properties: uniaxial neg.; $\omega = 1.640 - 1.652$, $\varepsilon = 1.638 - 1.649$, $\Delta = 0.003$ to 0.004.
Density: 3.07−3.16.
Crystallographic data: hexagonal, $P6_3/m$; $a \sim 9.48$ (Cl free), $c \sim 6.90$.

Prominent diffraction spacings, intensities to the base (10): (10—9) 2.75 and 2.84, (4—3) 3.45, 2.28 and 2.80, (3) 1.958 A.
Occurrences: Calcic-silicic metamorphic rocks.
Remarks: Inasmuch as wilkeite is an isomorphic variant intermediate between apatite and ellestadite (including hydroxylellestadite) its properties necessarily fall within such a range.

Mineral name: *voelckerite*
Compositional-structural name: oxyapatite
Formula: presumably $Ca_9A(PO_4)_6O_2$ and/or $Ca_{10}(PO_4)_5(ZO_4)O_2$ where A is $+3$ and Z is $+6$.
Occurrences: Isomorphic variants containing appreciable amounts of voelckerite seem to be rare, but occur in pegmatites.
Remarks: Data on any naturally occurring specimens are quite inadequate to permit indication of its properties, which are dependent necessarily on A and/or Z. An alternative premise is that the voelckerite formula can be written $Ca_{10}(PO_4)_6O$ $\square$, where $\square$ represents a vacancy in the structure. (This assumption is based on dubious evidence which includes inadequate analyses for water [164].)

Mineral name: (not known to occur as a mineral)
Compositional-structural name: octacalcium phosphate
Formula: $Ca_8H_2(PO_4)_6 \cdot 5H_2O$
Optical properties: biaxial neg.; $\gamma = 1.585$, $\beta = 1.583$, $\alpha = 1.576$, $2V = 50$ to $55°$.
Density: ~ 2.61.
Crystallographic data: triclinic, $P1$ or $P\bar{1}$; $a = 19.87$, $b = 9.63$, $c = 6.87$, $\alpha \sim 89°$, $\beta \sim 92°$, $\gamma \sim 109°$.
Prominent diffraction spacings, intensities to the base (10): (10) 18.6, 3.43, 2.84, 2.83 and 2.33, (9) 5.52, 3.74, 3.05, 2.77, 2.75, 2.67, 2.64 A and others.
Occurrences: Reported as a component of oral calculus.
Remarks: There has been much conjecture about this substance and its occurrences. The contribution of a dozen diffraction lines with intensities 9 or greater is most unusual [127], and agrees with A.S.T.M. No. 11—184 only with respect to the four most intense lines. High- to medium-intensity lines that are not similar to spacings of apatite are: 18.6, 9.46, 9.07 and 2.33 A.

References[1]

1. ADLER, H. H.: Infrared spectra of phosphate minerals: Splitting and frequency shifts associated with substitution of PO_4^{3-} for AsO_4^{3-} in mimetite ($Pb_5(AsO_4)_3Cl$). Amer. Mineral. **53**, 1740—1744 (1968).
2. AHRENS, L. H.: The use of ionization potentials. Part I. Ionic radii of the elements. Geochim. Cosmochim. Acta **2**, 155—169 (1952).
3. ALBEE, A. L., and A. A. CHODOS: Microprobe investigations on Apollo 11 samples. In: Proceedings of the Apollo 11 Lunar Science Conference, Houston, Texas, January 5—8, 1970. Vol. I, 135—157. (A. A. LEVINSON, ed.), New York: Pergamon Press, 1970.
4. ALTSCHULER, Z. S., S. BERMAN, and F. CUTTITTA: Rare earths in phosphorites—geochemistry and potential recovery. U.S. Geol. Surv., Prof. Paper **575-B**, 1—9 (1967).
5. ALTSCHULER, Z. S., R. S. CLARKE, JR., and E. J. YOUNG: Geochemistry of uranium in apatite and phosphorite. U.S. Geol. Surv., Prof. Paper **314-D**, 45—90 (1958).
6. AMES, L. L., JR.: The genesis of carbonate apatites. Econ. Geol. **54**, 829—841 (1959).
7. AMES, L. L., JR.: Some cation substitutions during the formation of phosphorite from calcite. Econ. Geol. **55**, 353—362 (1960).
8. ARMSTRONG, W. D.: Modification of the Willard-Winter method for fluorine determination. Jour. Amer. Chem. Soc. **55**, 1741—1742 (1933).
9. ARMSTRONG, W. D., and L. SINGER: Composition and constitution of the mineral phase of bone. Clinical Orthopaedics **38**, 179—190 (1965).
10. BANKS, E., M. GREENBLATT, and B. R. MCGARVEY: Electron spin resonance of CrO_4^{3-} in chloroapatite $Ca_5(PO_4)_3Cl$. Jour. Solid State Chem. **3**, 308—313 (1971).
11. BASSETT, C. A. L.: Electro-mechanical factors regulating bone architecture. Proc. European Symp. Calcified Tissues, 3rd 1965, 93—97 (1966).
12. BATES, R. L.: Geology of the Industrial Rocks and Minerals. New York: Harper & Bros. (1960) 441 pages.
13. BAUD, C. A., et S. SLATKINE: Données radiocristallographiques sur l'incorporation du fluor dans la substance minérale osseuse *in vivo*. C. R. Acad. Sci. Paris **255**, 1801—1802 (1962).
14. BAUMHAUER, H.: Über die Brechungsexponenten und die Doppelbrechung des Apatits von verschiedenen Fundorten. Zeits. Kristallogr. **45**, 555—568 (1908).
14a. BELINKO, G. DE: Conditions océanographiques de la genèse des phosphates sédimentaires. C. R. Acad. Sci., Paris **269**, sér. D, 875—877 (1969); also 935—938.
15. BHATNAGAR, V. M.: The melting points of synthetic apatites. Mineral. Mag. **37**, 527—528 (1969).
16. BHATNAGAR, V. M.: The cell parameters of synthetic fluorapatite, $Ca_{10}(PO_4)_6F_2$. Rev. Roumaine Chim. **15**, 87—89 (1970).
17. BHATNAGAR, V. M.: The cell parameters of synthetic calcium apatites. Rev. Romaine Chim. **15**, 1735—1739 (1970).
18. BHATNAGAR, V. M.: Preparation and X-ray powder diffraction patterns of lead apatites. Chem. Ind. (London) **48**, 1538—1539 (1970).

---

[1] Titles enclosed in brackets are translations.

19. Biggar, G. M.: Apatite; the correlation of refractive index with substitution. Indian Mineralogist **8**, 35—45 (1968).
20. Bonel, G., et G. Montel: Sur une nouvelle apatite carbonatée synthétique. C. R. Acad. Sci. Paris **258**, 923—926 (1964).
21. Borgström, L. H.: Syntetisk NaCl-apatit. Finska Kemi. Meddel. **2**, 51—54 (1932).
22. Brindley, G. W., and D. Fancher: Kaolinite defect structures; possible relation to allophanes. Proc. Internat. Clay Conf., Tokyo, Japan **2**, 29—34 (1970).
23. Brögger, W.C., und H. Bäckström: Über den Dahllit ein neues Mineral von Ödegården, Bamle, Norwegen. Öfversigt K. Vet.-Akad. Förhandl. [Stockholm] **45**, 493—496 (1888).
24. Brophy, G. P., and T. M. Hatch: Recrystallization of fossil horse teeth. Amer. Mineral. **47**, 1174—1180 (1962).
25. Brophy, G. P., and T. J. Nash: Composition, infrared, and X-ray analysis of fossil bone. Amer. Mineral. **53**, 445—454 (1968).
26. Brown, W. E., J. R. Lehr, J. P. Smith, and A. W. Frazier: Crystallography of octacalcium phosphate. Jour. Amer. Chem. Soc. **79**, 5318 (1957).
27. Bukanov, V. V.: [Apatite from alpine-type veins in the Polar Urals.] Zap. Vses. Mineral. Obshch. **90**, 591—598 (1961).
28. Bushinsky, G. I.: [Old Phosphates of Asia and their Genesis.] Akad. Nauk SSSR, Geol. Inst., Trans. **149**, 1—192 (33 plates). Izd. "Nauka", Moscow, 1966.
29. Bushinsky, G. I.: [Phosphoria Formation.] Akad. Nauk SSSR, Geol. Inst., Trans. **201**, 1—103 (20 plates). Moscow: Izd. "Nauka" (1969).
30. Cady, J. G., W. L. Hill, E. V. Miller, and R. M. Magness: Occurrence of beta tricalcium phosphate in northern Mexico. Amer. Mineral. **37**, 180—183 (1952).
31. Capen, C. C., C. R. Cole, and J. W. Hibbs: The pathology of hypervitaminosis D in cattle. Path. Vet. **3**, 350—378 (1966).
32. Carlström, D.: Mineralogical carbonate containing apatites. Unpublished manuscript presented at the Internat. Symp. on Structural Properties of Hydroxyapatite and Related Compounds, Gaithersburg, Maryland, Sept. 12—14, 1968.
33. Carobbi, G.: Celestina e apatite di stronzio contenenti piccole quantita di mercurio. Accad. Naz. Lincei Atti [ser. 8] **8**, 87—93 (1950).
34. Carpenter, R.: Factors controlling the marine geochemistry of fluorine. Geochim. Cosmochim. Acta **33**, 1153—1167 (1969).
35. Clark, S. P., Jr. (ed.): Handbook of Physical Constants. New York: Geol. Soc. Amer., Mem. **97** (1966) 587 pages.
36. Clarke, F. W.: The Data of Geochemistry. U.S. Geol Surv. Bull. **770** (1924) 841 pages.
37. Cockbain, A. G.: The crystal chemistry of apatites. Mineral Mag. **36**, 654—660 (1968).
38. Collin, R. L.: Strontium-calcium hydroxyapatite solid solutions precipitated from basic, aqueous solutions. J. Amer. Chem. Soc. **82**, 5067—5069 (1960).
39. Collins, R. M.: The mining and preparation of Florida land pebble phosphate. Proc. Fertil. Sci., No. **9**, 1—26 (1950).
40. Cooray, P. G.: A carbonate-bearing fluor-chlor-hydroxyapatite from Matale, Ceylon. Amer. Mineral. **55**, 2038—2041 (1970).
41. Cruft, E. F., C. O. Ingamells, and J. Muysson: Chemical analysis and stoichiometry of apatite. Geochim. Cosmochim. Acta **29**, 581—597 (1965).
42. Dana, E. S., and W. E. Ford: A Textbook of Mineralogy. 4th ed., New York: John Wiley & Sons (1932) 851 pages.
42a. D'Anglejan, B. F.: Origin of marine phosphorites off Baja California, Mexico. Marine Geol. **5**, 15—44 (1967).
43. Davis, K. A.: The phosphate deposits of Eastern Province, Uganda. Econ. Geol. **42**, 137—146 (1947).
44. Deans, T.: Economic mineralogy of African carbonatites. In: Carbonatites, O. F. Tuttle and J. Gittins (eds.). New York: Interscience Publishers, 1966. Pages 385—413.

45. DEANS, T., and H. C. G. VINCENT: Francolite from sedimentary ironstones of the Coal Measures. Mineral. Mag. **25**, 135—139 (1938).
46. DEER, W. A., R. A. HOWIE, and J. ZUSSMAN: Rock-Forming Minerals, vol. **5**, 323—338. London: Longmans, Green & Co., 1962.
47. DELITSINA, L. V., and B. N. MELENT'EV: [Coexistence of liquid phases at high temperatures. Apatite-nepheline-villiaumite system.] Dokl. Akad. Nauk SSSR **188**, 431—433 (1969). [Transl. **188**, 185—187.]
47a. DIBDIN, G. H.: The stability of water in human dental enamel studied by proton nuclear magnetic resonance. Arch. oral Biol. **17**, 433—437 (1972).
48. DORFMAN, M. D.: [Francolite from Khibina tundras.] Izvest. Karel'sk. i Kol'sk Filial. Akad. Nauk SSSR **1958**, No. 4, 32—39 (1958).
49. DRAUS, F. J., F. L. MIKLOS, and S. W. LEUNG: Carbonic anhydrase in a bovine submaxillary gland extract. Jour. Dental Res. **41**, 497 (1962).
49a. DUFF, E. J.: Orthophosphates. Part IV. Stability relationship of orthophosphates within the system CaO-P_2O_5-H_2O and CaF_2-CaO-P_2O_5-H_2O under aqueous conditions. Jour. Chem. Soc. **1971 A**, 921—926 (1971).
50. EAKLE, A. S., and A. F. ROGERS: Wilkeite, a new mineral of the apatite group, and okenite, its alteration product, from Southern California. Amer. Jour. Sci. [4] **37**, 262—267 (1914).
51. ELLIOTT, J. C.: Recent progress in the chemistry, crystal chemistry and structure of the apatites. Calc. Tissue Res. **3**, 293—307 (1969).
52. ELLISON, S.: The composition of conodonts. Jour. Paleo. **18**, 133—140 (1944).
53. EMERSON, W. H., and E. E. FISCHER: The infra-red absorption spectra of carbonate in calcified tissues. Arch. oral Biol. **7**, 671—683 (1962).
54. ENGEL, G.: Einige Cadmiumapatite sowie die Verbindungen Cd_2XO_4F mit $X = P$, As und V. Zeits. anorg. allg. Chem. **378**, 49—61 (1970).
55. ENGEL, G.: Hydrothermalsynthese von Bleihydroxylapatiten $Pb_5(XO_4)_3OH$ mit $X = P$, As, V. Naturwiss. **57**, 355 (1970).
56. ENNEVER, J., J. J. VOGEL, and J. L. STRECKFUSS: Synthetic medium for calcification of *Bacterionema matruchotii*. Jour. Dental Res. **50**, 1327—1330 (1971).
57. FEDEROV, N. F., I. F. ANDREEV, A. M. SHEVJAKOV, and T. A. TUNIK: [Silicoapatites of Sr and Nb.] Izvest. Akad. Nauk SSSR, Neorg. Mater. **6**, 2018—2021 (1970).
58. FELSCHE, J.: A new cerium (III) orthosilicate with the apatite structure. Naturwiss. **56**, 325—326 (1969).
59. FINE, B. S., J. W. BERKOW, and S. FINE: Corneal calcification. Science **162**, 129—130 (1968).
60. FISCHER, R. B., and C. E. RING: Quantitative infrared analysis of apatite mixtures. Anal. Chem. **29**, 431—434 (1957).
61. FISHER, D. J.: Pegmatite phosphates and their problems. Amer. Mineral. **43**, 181—207 and 609—610 (1958).
61a. FISHER, D. J.: Morinite-apatite-whitlockite. Amer. Mineral. **45**, 645—667 (1960).
62. FISHER, D. J., and D. MCCONNELL: Aluminium-rich apatite. Science **164**, 551—553 (1969).
63. FLEISCH, H., and W. F. NEUMAN: Mechanism of calcification: role of collagen, polyphosphates, and phosphatase. Amer. Jour. Physiol. **200**, 1296—1300 (1961).
63a. FLEISCHER, M.: Glossary of Mineral Species. Bowie, Maryland: Mineralogical Record, Inc., 1971, 1—103.
64. FLEISCHER, M., and Z. S. ALTSCHULER: The relationship of rare-earth composition to geological environment. Geochem. Cosmochim. Acta **33**, 725—732 (1969).
65. FORD, W. E.: [see E. S. DANA].
66. FOREMAN, D. W., JR.: Neutron and X-ray diffraction study of $Ca_3Al_2(O_4D_4)_3$, a garnetoid. Jour. Chem. Phys. **48**, 3037—3041 (1968).

67. Foshag, W. F.: Mineralogy and geology of Cerro Mercado, Durango, Mexico. U.S. Natl. Mus., Proc. **74**, art. 23 (1929) 27 pages.
68. Francis, M. D., and N. C. Webb: Hydroxyapatite formation from a hydrated calcium monohydrogen phosphate precursor. Calc. Tissue Res. **6**, 335–342 (1971).
69. Frondel, C.: Mineralogy of the calcium phosphates in insular phosphate rock. Amer. Mineral. **28**, 215–232 (1943).
70. Frondel, C.: [see Palache].
71. Fuchs, L. H.: Fluorapatite and other accessory minerals in Apollo 11 rocks. In: Proceedings of the Apollo 11 Lunar Science Conference, Houston, Texas, January 5–8, 1970. Vol. I, 475–479. (A. A. Levinson, ed.), New York: Pergamon Press, 1970.
72. Geijer, P.: Internal features of the apatite-bearing magnetite ores. Sveriges Geol. Unders. (ser. C) **624**, 1–32 (1967).
72a. Girault, J.: Genèse et géochimie de l'apatite et de la calcite dans les roches liées au complexe carbonatitique et hyperalcalin d'Oka (Canada). Bull. Soc. franç. Minéral. Cristal. **89**, 496–513 (1966).
72b. Gilinskaya, L. G., M. Y. Shcherbakova, and Y. N. Zanin: [Carbon in the structure of apatite according to electron paramagnetic resonance data.] Kristallografiya **15**, 1164–1167 (1970) [Transl. **15**, 1016–1019 (1971)].
73. Gluskoter, H. J., L. H. Pierard, and H. W. Pfefferkorn: Apatite pertifications in Pennsylvanian shales of Illinois. Jour. Sed. Petrology **40**, 1363–1366 (1970).
74. Goldschmidt, V.: Atlas der Kristallformen. Vol. 1, 248 pages. Heidelberg: Carl Winters Universitätsbuchhandlung, 1913, 244 plates.
75. Goldsmith, J. R., D. L. Graf, and O. I. Joensuu: The occurrence of magnesian calcites in nature. Geochim. Cosmochim. Acta **7**, 212–230 (1955).
75a. Gopal, R., and C. Calvo: Structural relationship of whitlockite and $\beta Ca_3(PO_4)_2$. Nature Phys. Sci. **237**, 30–32 (1972).
76. Grasselly, G.: On the phosphorus-bearing mineral of the manganese oxide ore deposits of Eplény and Urkút. Acta Univ. Szegediensis Miner.-Petrogr. **18**, 73–83 (1968).
77. Greenblatt, M., E. Banks, and B. Post: The crystal structures of the spodiosite analogs, Ca_2CrO_4Cl and Ca_2PO_4Cl. Acta Cryst. **23**, 166–171 (1967).
78. Grisafe, D. A., and F. A. Hummel: Crystal chemistry and color in apatites containing cobalt, nickel, and rare-earth ions. Amer. Mineral. **55**, 1131–1145 (1970).
79. Grisafe, D. A., and F. A. Hummel: Pentavalent ion substitutions in the apatite structure. A. Crystal chemistry. J. Solid State Chem. **2**, 160–166 (1970).
80. Gruner, J. W., and D. McConnell: The problem of the carbonate apatites. The structure of francolite. Zeits. Kristallogr. **97 A**, 208–215 (1937).
81. Gruner, J. W., D. McConnell, and W. D. Armstrong: The relationship between crystal structure and chemical composition of enamel and dentin. J. Biol. Chem. **121**, 771–781 (1937).
82. Gulbrandsen, R. A.: Chemical composition of phosphorites of the Phosphoria Formation. Geochim. Cosmochim. Acta **30**, 769–778 (1966).
82a. Gulbrandsen, R. A., J. R. Kramer, L. B. Beatty, and R. E. Mays: Carbonate-bearing apatite from Faraday Township, Ontario, Canada. Amer. Mineral. **51**, 819–824 (1966).
83. Habashi, F.: Correlation between the uranium content of marine phosphates and other rock constituents. Econ. Geol. **57**, 1081–1084 (1962).
83a. Habashi, F.: Uranium in phosphate rock. Montana Bur. Mines Geol., Spl. Publ. **52**, 1–33 (1970).
84. Haberlandt, H.: Spektralanalytische Untersuchungen und Lumineszenzbeobachtungen an Fluoriten und Apatiten. Sitzungsber. Akad. Wiss. Wien, math.-natur. Kl., Abt. II a, **147**, 137–150 (1938).

85. HAMMEN, C. S., D. P. HANLON, and S. C. LUM: Oxidative metabolism of *Lingula*. Comp. Biochem. Physiol. **5**, 185—191 (1962).
86. HARADA, K., K. NAGASHIMA, K. NAKAO, and A. KATO: Hydroxylellestadite, a new apatite from Chichibu mine, Saitama Prefecture, Japan. Amer. Mineral. **56**, 1507—1518 (1971).
87. HARRIS, R. A., D. F. DAVIDSON, and B. P. ARNOLD: Bibliography of the Geology of the Western Phosphate Field. U.S. Geol. Surv., Bull. **1018**, 1—89 (1954).
88. HASS, W. H., and M. L. LINDBERG: Orientation of the crystal units of conodonts. Jour. Paleo. **20**, 501—504 (1946).
89. HAÜY, R. J.: Traité de Minéralogie, vol. 1, Paris: Bachelier, 1822, p. 503.
90. HAUSEN, H.: Die Apatite, deren chemische Zusammensetzung und ihr Verhältnis zu den physikalischen und morphologischen Eigenschaften. Acta Acad. Åbo, math.-phys. **5**, [No. 3] (1929) 62 pp.
90a. HAYASHI, S.: [Analysis of the crystal structure of fluorapatite by nuclear magnetic resonance.] J. Chem. Soc. Japan **81**, 540—542 (1960). [Japanese with English summary.]
91. HENDRICKS, S. B., and W. L. HILL: The inorganic constitution of bone. Science **96**, 255—257 (1942).
92. HENDRICKS, S. B., M. E. JEFFERSON, and V. M. MOSLEY: The crystal structures of some natural and synthetic apatite-like substances. Zeits. Kristallogr. **81**, 352—369 (1932).
93. HENTSCHEL, H.: Röntgenographische Untersuchungen am Apatit. Centralb. Mineral. Geol. Paleont. **1923**, 609—626 (1923).
94. HODGE, H. C.: Hardness tests on teeth. Jour. Dental Res. **15**, 271—279 (1936).
95. HOFFMAN, V., und Z. TRDLIČKA: Über ein Carbonat-Apatit (Francolith) von Kutná Hora. Acta Univ. Carolinae, Geol. No. **3**, 195—202 (1967).
96. HOFFMANN, J.: Über Ionen- und Atomfärbungen künstlich hergestellter und natürlicher Apatite. Chemie der Erde **11**, 552—575 (1938).
97. HOUNSLOW, A. W., and G. Y. CHAO: Monoclinic chlorapatite from Ontario. Canadian Mineral. **10**, 252—259 (1970).
98. HOWIE, R. A.: A pyroxene granulite from Hitterö, southwest Norway. Indian Geophys. Union, Krishnan vol. 297—307 (1964).
99. HUTCHINSON, G. E.: Survey of contemporary knowledge of biochemistry. 3. The biogeochemistry of vertebrate excretion. Bull. Amer. Mus. Nat. Hist. **96**, 554 p. (1950).
100. HUTTON, C. O., and F. T. SEELYE: Francolite, a carbonate-apatite from Milburn, Otago. Roy. Soc. N. Zealand, Trans. **72**, 191—198 (1942).
101. IGELSRUD, I., J. CHOCHOLAK, and E. F. STEPHAN et al.: Recovery of uranium from phosphate rock by the Battelle monocalcium phosphate process. U.S. Atomic Energy Comm. BMI-JDS-201 (1949) 99 pages.
102. ITO, J.: Silicate apatites and oxyapatites. Amer. Mineral. **53**, 890—907 (1968).
103. JAFFE, E. B.: Abstracts of the literature on synthesis of apatites and some related phosphates. U.S. Geol. Surv., Circular **135** (1951) 78 pages.
104. JAFFE, H. W., and V. J. MOLINSKI: Spencite, the yttrium analogue of tritomite from Sussex County, New Jersey. Amer. Mineral. **47**, 9—25 (1962).
105. JOHNSON, P. D.: Oxygen-dominated lattices. In: Luminescence of Inorganic Solids, (PAUL GOLDBERG, ed.). New York: Academic Press, 1966. Pages 287—336.
106. JOHNSON, P. D., J. S. PRENER, and J. D. KINGSLEY: Apatite: origin of blue color. Science **141**, 1179—1180 (1963).
107. JOHNSON, W.: Two synthetic compounds containing chromium in different valency states. Mineral. Mag. **32**, 408—411 (1960).
108. KANESHIMA, K.: [Geochemistry of the phosphate rocks in the Ryukyu Islands. (1) Content and behavior of zinc in the phosphate rocks.] Nippon Kagaku Zasshi **83**, 1007—1011 (1962) [Japanese, English summary].

109. KATZ, S., C. W. BECK, and J. C. MUHLER: Crystallographic evaluation of enamel from carious and noncarious teeth. Jour. Dental Res. **48**, 1280—1283 (1969).
110. KIND, A.: Der magmatische Apatit, seine chemische Zusammensetzung und seine physikalischen Eigenschaften. Chemie der Erde **12**, 50—81 (1938).
111. KISLOVSKIY, K. D., and R. G. KNUBOVETS: [Sensitivity of infrared spectra of single apatite crystals to isomorphism.] Dokl. Akad. Sci. U.S.S.R., Earth Sci. Sect. **179**, 138—141 (1968). [Transl. **179**, 1432—1435.]
112. KLEE, W. E.: The vibrational spectra of the phosphate ions in fluorapatite. Zeits. Kristallogr. **131**, 95—102 (1970).
113. KLEE, W. E., and G. ENGEL: I. R. spectra of the phosphate ions in various apatites. J. Inorg. Nucl. Chem. **32**, 1837—1843 (1970).
114. KLEMENT, R.: Der Carbonatgehalt der anorganischen Knochensubstanz und ihre Synthese. Ber. Deutschen Chem. Gesellschaft **69**, 2232—2238 (1936).
115. KLEMENT, R.: Isomorpher Ersatz des Phosphors in Apatiten durch Silicium und Schwefel. Naturwiss. **27**, 57—58 (1939).
116. KLEMENT, R.: Natrium-Calcium-sulfapatite $Na_6Ca_4(SO_4)_6F_2$. Naturwiss. **27**, 568 (1939).
117. KLEMENT, R., und H. HASELBECK: Apatite und Wagnerite zweiwertiger Metalle. Zeits. anorg. allg. Chem. **336**, 113—128 (1965).
118. KOLODNY, Y., and I. R. KAPLAN: Carbon and oxygen isotopes in apatite CO_2 and co-existing calcite from sedimentary phosphorite. Jour. Sed. Petrology **40**, 954—959 (1970).
119. KREIDLER, E. R., and F. A. HUMMEL: The crystal chemistry of apatite: structure fields of fluor- and chlorapatite. Amer. Mineral. **55**, 170—184 (1970).
120. LAGERGREN, C.: Biophysical investigation of urinary calculi. Acta Radiol., Supplement **133**, 71 pages (1965).
121. LARGENT, E. J.: Fluorosis, the Health Aspects of Fluorine Compounds. Columbus: Ohio State Univ. Press, 140 pages (1961).
122. LARSEN, E. S., and E. V. SHANNON: Two phosphates from Dehrn; dehrnite and crandallite. Amer. Mineral. **15**, 303—306 (1930).
123. LARSEN, E. S., and E. V. SHANNON: The minerals of the phosphate nodules from near Fairfield, Utah. Amer. Mineral. **15**, 307—337 (1930).
124. LARSEN, E. S., JR., M. H. FLETCHER, and E. A. CISNEY: Strontian apatite. Amer. Mineral. **37**, 656—658 (1952).
125. LEGEROS, R. Z., J. P. LEGEROS, O. R. TRAUTZ, and E. KLEIN: Spectral properties of carbonate in carbonate-containing apatites. Developments. Appl. Spectral **7 B**, 3—12 (1970).
126. LEGEROS, R. Z., O. R. TRAUTZ, E. KLEIN, and J. P. LEGEROS: Two types of carbonate substitution in the apatite structure. Experientia **25**, 5—7 (1969).
127. LEHR, J. R., E. H. BROWN, A. W. FRAZIER, J. P. SMITH, and R. D. THRASHER: Crystallographic properties of fertilizer compounds. Tennessee Valley Authority, Chem. Engr. Bull. **6**, 166 pages (1967).
128. LEHR J. R., G. H. MCCLELLAN, J. P. SMITH, and A. W. FRAZIER: Characterization of apatites in commercial phosphate rocks. Colloque Intern. Phosphates Minéraux Solides, Toulouse 1967. Paris: Masson, 1968, 184 pages; pp. 29—44.
129. LEVINSON, A. A., and S. R. TAYLOR: Moon Rocks and Minerals. Scientific results of the study of Apollo 11 lunar samples with preliminary data on Apollo 12 samples. New York: Pergamon Press, 1971, 222 pages.
130. LEVITT, S. R., and R. A. CONDRATE, SR.: The vibrational spectra of lead apatites. Amer. Mineral **55**, 1562—1575 (1970).
131. LINDBERG, M. L., and B. INGRAM: Rare-earth silicatian apatite from the Adirondack Mountains, New York. U.S. Geol. Surv., Prof. Paper **501-B**, 64—65 (1964).

132. LITTLE, M. F.: Studies on the inorganic carbon dioxide component of human enamel. II. The effect of acid on enamel CO_2. Jour. Dental Res. **40**, 903—914 (1961).
133. LITTLE, M. F., and F. S. CASCIANI: The nature of water in sound human enamel. A preliminary study. Arch. oral Biol. **11**, 565—571 (1966).
134. LONSDALE, K.: Human stones. Science **159**, 1199—1207 (1968).
135. LOWENSTAM, H. A.: Biologic problems relating to the composition and diagenesis of sediments. In: The Earth Sciences, (T. W. DONNELLY, ed.). Houston: Rice Univ., Semicentennial Publs. 1963, 137—195.
136. MCCLELLAN, G. H., and J. R. LEHR: Crystal chemical investigation of natural apatites. Amer. Mineral. **54**, 1374—1391 (1969).
137. MCCONNELL, D.: The substitution of SiO_4 and SO_4 groups for PO_4 groups in the apatite structure; ellestadite, the end member. Amer. Mineral. **22**, 977—986 (1937).
138. MCCONNELL, D.: A structural investigation of the isomorphism of the apatite group. Amer. Mineral. **23**, 1—19 (1938).
139. MCCONNELL, D.: The petrography of rock phosphates. Jour. Geol. **58**, 16—23 (1950).
140. MCCONNELL, D.: The problem of the carbonate apatites. IV. Structural substitutions involving CO_3 and OH. Bull. Soc. franç. Minéral. Crist. **75**, 428—445 (1952).
141. MCCONNELL, D.: The crystal chemistry of francolite and its relationship to calcified animal tissues. In: Metabolic Interrelations, Trans. 4th Conference, (E. C. REIFENSTEIN, ed.). New York: Josiah Macy, Jr. Foundation, 1952, pages 169—184.
142. MCCONNELL, D.: Radioactivity of phosphatic sediments. Econ. Geol. **48**, 147—148 (1953).
143. MCCONNELL, D.: The apatitelike mineral of sediments. Econ. Geol. **53**, 110—111 (1958).
144. MCCONNELL, D.: The problem of the carbonate apatites. Econ. Geol. **54**, 749—751 (1959).
145. MCCONNELL, D.: The crystal chemistry of dahllite. Amer. Mineral. **45**, 209—216 (1960).
146. MCCONNELL, D.: Carbonate in apatites. Science **134**, 213 (1961).
147. MCCONNELL, D.: The birefringence of carbonate apatites. Mineral. Mag. **33**, 65—66 (1962).
148. MCCONNELL, D.: Dating of fossil bones by the fluorine method. Science **136**, 241—244 (1962).
149. MCCONNELL, D.: Thermocrystallization of richellite to produce a lazulite structure (calcium lipscombite). Amer. Mineral. **48**, 300—307 (1963).
150. MCCONNELL, D.: Inorganic constituents in the shell of the living brachiopod *Lingula*. Geol. Soc. Amer. Bull. **74**, 363—364 (1963).
151. MCCONNELL, D.: Refringence of garnets and hydrogarnets. Canadian Mineral. **8**, 11—22 (1964).
152. MCCONNELL, D.: Deficiency of phosphate ions in apatite. Naturwiss. **52**, 183 (1965).
153. MCCONNELL, D.: Crystal chemistry of hydroxyapatite: its relation to bone mineral. Arch. oral Biol. **10**, 421—431 (1965).
154. MCCONNELL, D.: Precipitation of phosphates in sea water. Econ. Geol. **60**, 1059—1062 (1965).
155. MCCONNELL, D.: Calculation of the unit-cell volume of a complex mineral structure. Zeits. Kristallogr. **123**, 58—66 (1966).
156. MCCONNELL, D.: Crystal chemical calculations. Geochim. Cosmochim. Acta **31**, 1479—1487 (1967).
157. MCCONNELL, D.: Infrared absorption of carbonate apatite. Science **155**, 607 (1967).
158. MCCONNELL, D.: Crystal chemistry of bone mineral: Hydrated carbonate apatites. Amer. Mineral. **55**, 1659—1669 (1970).
159. MCCONNELL, D.: Crystal chemistry of phosphorite. Econ. Geol. **66**, 1085—1086 (1971).
160. MCCONNELL, D., and D. W. FOREMAN, JR.: The properties and structure of $Ca_{10}(PO_4)_6(OH)_2$; its relation to tin (II) apatite. Canadian Mineral. **8**, 431—436 (1966).
161. MCCONNELL, D., and D. W. FOREMAN, JR.: Crystallochemical differences between carious and noncarious dental enamel. Program and Abstracts, 49th Gen. Session, Internat. Asn. Dental Res., Chicago, March 18—21, 1971, page 256.

161a. McConnell, D., and D. W. Foreman, Jr.: Texture and composition of bone. Science **172**, 971–972 (1971).
162. McConnell, D., W. J. Frajola, and D. W. Deamer: Relation between the inorganic chemistry and biochemistry of bone mineralization. Science **133**, 281–282 (1961).
163. McConnell, D., and J. W. Gruner: The problem of the carbonate apatites. III-Carbonate-apatite from Magnet Cove, Arkansas. Amer. Mineral. **25**, 157–167 (1940).
164. McConnell, D., and M. H. Hey: The oxyapatite (voelckerite) problem. Mineral. Mag. **37**, 301–303 (1969).
165. McConnell, D., and J. Murdoch: Crystal chemistry of scawtite. Amer. Mineral. **43**, 498–502 (1958).
166. McConnell, D., and J. Murdoch: Crystal chemistry of ettringite. Mineral. Mag. **33**, 59–64 (1962).
167. McKeown, F. A., and H. Klemic: Rare-earth-bearing apatite at Mineville, Essex County, New York. U.S. Geol. Surv., Bull. **1046-B**, 9–23 (1957).
168. McLean, F. C., and A. M. Budy: Radiation, Isotopes, and Bone. New York: Academic Press, 1964, 216 pages.
169. McLean, F. C., and M. R. Urist: Bone, Fundamentals of the Physiology of Skeletal Tissue. 3rd ed., Chicago: Univ. Chicago Press, 1968, 314 pages.
170. Märk, E., M. Pahl, and D. Märk: Fission-Track-Alter von Durango-Apatit, Mexiko. Contr. Mineral. Petrol **32**, 147–148 (1971).
171. Martens C. S., and R. C. Harriss: Inhibition of apatite precipitation in the marine environment by magnesium ions. Geochim. Cosmochim. Acta **34**, 621–625 (1970).
172. Mason, B.: Principles of Geochemistry, 3rd. ed., New York: John Wiley & Sons, 1966, 329 pp.
173. Mayer, I., and V. Makogon-Loewy: Eu(II)-hydroxyapatite. Preparation and crystal data. Israel Jour. Chem. **7**, 717–719 (1969).
173a. Mazor, E.: Notes concerning the geochemistry of phosphorus, fluorine, uranium and radium in some marine rocks in Israel. Israel Jour. Earth-Sci. **12**, 41–52 (1963).
174. Mehmel, M.: Über die Struktur des Apatits I. Zeits. Kristallogr. **75**, 323–331 (1930); *also* Beziehungen zwischen Kristallstruktur und chemischer Formel des Apatits. Zeits. physik. Chemie **15-B**, 223–241 (1931).
175. Merker, L., G. Engel, H. Wondratschek, and J. Ito: Lead ions and empty halide sites in apatites. Amer. Mineral. **55**, 1435–1437.
176. Minguzzi, C.: Apatiti sintetiche con cromo trivalente ed esavalente. Periodico Mineral. (Rome) **12**, 343–378 (1941).
177. Mitchell, L., G. T. Faust, S. B. Henricks, and D. S. Reynolds: The mineralogy and genesis of hydroxylapatite. Amer. Mineral **28**, 356–371 (1943).
177a. Mitchell, R. S., and E. H. McGavock: Apatite from the Morefield pegmatite, Amelia County, Virginia. Rocks and Minerals **35**, 553–555 (1960).
178. Moore, R. E., and W. Eitel: A borosilicate of the apatite group. Naturwiss. **44**, 259 (1957).
178a. Moreno, E. C., T. M. Gregory, and W. E. Brown: Preparation and solubility of hydroxyapatite. Natl. Bur. Standards. Jour. Res. **72 A**, 773–782 (1968).
179. Morton, R. D., and E. J. Catanzaro: Stable chlorine isotope abundances in apatites from Ødegårdens verk, Norway. Norsk Geol. Tidsskr. **44**, 307–313 (1964).
180. Náray-Szabó, St.: The structure of apatite (CaF)Ca$_4$(PO$_4$)$_3$. Zeits. Kristallogr. **75**, 387–398 (1930).
180a. Nathan, Y., et J. Lucas: Synthèse de l'apatite à partir du gypse; application au problème de la formation des apatites carbonatées par précipitation directe. Chem. Geol. **9**, 99–112 (1972).

181. Nordin, B. E. C., M. Bluhm, and J. MacGregor: In vitro and in vivo studies with bone-seeking isotopes. In: Radio-Isotopes and Bone, (P. Lacroix and A. M. Budy, eds.). Oxford: Blackwell, 1962, pages 105—126.
182. Oleksynowa, K.: [The mineralogy and chemistry of the phosphate rock from Chalupki.] Polska Akad. Nauk. Arch. Mineral. **23**, 215—264 [265—270 in English] (1962).
183. Omori, K., and H. Konno: A new yttrian apatite enclosed in quartz from Naegi, Gifu Prefecture, Japan. Amer. Mineral. **47**, 1191—1195 (1962).
184. Ørvig, T.: Evolution of some calcified tissues in early vertebrates. In A. E. W. Miles, ed., Structural and Chemical Organization of Teeth, vol. **I**. London: Academic Press, 1967, 525 pages.
185. Ostaszewicz, E.: On the fluorescence of Sb- and Mn-activated calcium halophosphates. Acta Phys. Polon. **19**, 421—442 (1960).
185a. Palache, C., H. Berman, and C. Frondel: The System of Mineralogy, Vol. **II**, 7th ed. New York: John Wiley & Sons, 1951, 1124 pages.
186. Parker, R. B., and H. Toots: Minor elements in fossil bone. Geol. Soc. Amer. Bull. **81**, 925—932 (1970).
186a. Parker, R. J.: The petrography and major element geochemistry of phosphorite nodule deposits on the Agulhas Bank, South Africa. S. Africa Nat. Com. Oceanogr. Res., Bull. **2**, 1—94 (1971).
187. Peacock, J. D., and K. Taylor: Uraniferous collophane in the Carboniferous Limestone of Derbyshire and Yorkshire. Geol. Surv. Gr. Britain Bull. **25**, 19—32 (1966).
187a. Phakey, P. P., and J. R. Leonard: Dislocations and fault surfaces in natural apatite. Jour. Appl. Cryst. **3**, 38—44 (1970).
188. Pigorini, B., e F. Veniale: L'apatite accessoria nelle diverse facies litologiche delle formazioni granitoidi della Val Sessera. Rend. Soc. Min. Italiana **24**, 283—312 (1968).
189. Poitevin, E.: A new Canadian occurrence of phosphorites from near François Lake, British Columbia. Canada Dept. Mines, Bull. **46**, 2—12 (1927).
190. Portnov, A. M., V. T. Dubinchuk, and T. I. Stolyarova: A natural rare-earth oxyapatite. Dokl. Akad. Nauk SSSR **192**, 881—884 (1970) [transl. **192**, 114—116].
191. Portnov, A. M., and B. S. Gorobets: [Luminescence of apatite from different rock types.] Dokl. Akad. Nauk SSSR **184**, 199—202 (1969) [transl. **184**, 110—113].
192. Prener, J. S.: The growth and crystallographic properties of calcium fluor- and chlorapatite crystals. Jour. Electrochem. Soc. **114**, 77—83 (1967).
193. Prien, E. L., and C. Frondel: Studies in urolithiasis: I. The composition of urinary calculi. Jour. Urol. **57**, 949—991 (1947).
194. Quensel, P.: The paragenesis of the Varuträsk pegmatite. Arkiv Mineral. Geol. (Stockholm) **2**, 54—61 (1956).
195. Reeves, M. J., and T. A. K. Saadi: Factors controlling the deposition of some phosphate bearing strata from Jordan. Econ. Geol. **66**, 451—465 (1971).
195a. Reif, W.-E.: Zur Genese des Muschelkalk-Keuper-Grenzbonebeds in Südwestdeutschland. Neues Jahrb. Paläont. Abh. **139**, 369—404 (1971).
195b. Retief, D. H., P. E. Cleaton-Jones, J. Turkstra, and W. J. de Wet: The quantitative analysis of sixteen elements in normal human enamel and dentine by neutron activation analysis and high-resolution gamma-spectroscopy. Arch. oral Biol. **16**, 1257—1267 (1971).
196. Rizzo, A. A., D. B. Scott, and H. A. Bladen: Calcification of oral bacteria. Ann. New York Acad. Sci. **109**, 14—22 (1963).
197. Rogers, A. F.: A new locality for voelckerite and its validity as a mineral species. Mineral. Mag. **17**, 155—162 (1914).
197a. Rooney, T. P., and P. F. Kerr: Mineralogic nature and origin of phosphorite, Beaufort County, North Carolina. Geol. Soc. Amer. Bull. **78**, 731—748 (1967).

198. SANDELL, E. B., M. H. HEY, and D. McCONNELL: The composition of francolite. Mineral. Mag. **25**, 395—401 (1939).
199. SCHIBLER, D., and H. FLEISCH: Inhibition of skin calcification (calciphylaxis) by polyphosphates. Experientia **22**, 367—369 (1966).
200. SCHNEIDER, W.: Caracolit, das $Na_3Pb_2(SO_4)_3Cl$ mit Apatitstruktur. Neues Jahrb., Mineral. Monatsh. **1967**, 284—289 (1967).
201. SHANNON, R. D., and C. T. PREWITT: Effective ionic radii in oxides and fluorides. Acta Crystallogr. **B 25**, 925—946 (1969).
202. SHCHERBOKOVA, M. Y., L. G. GILINSKAYA, and A. A. GODOVIKOV: [Paramagnetic centers in apatite.] Kristallografiya **13**, 353—356 (1968) [transl. **13**, 289—290].
203. SHELDON, R. P.: Geochemistry of uranium in phosphorites and black shales of the Phosphoria Formation. U.S. Geol. Surv., Bull **1084-D**, 83—113 (1959).
204. SILVERMAN, S. R., R. K. FUYAT, and J. D. WEISER: Quantitative determination of calcite associated with carbonate-bearing apatite. Amer. Mineral. **37**, 211—222 (1952).
205. SIMPSON, D. R.: The nature of alkali carbonate apatites. Amer. Mineral. **49**, 363—376 (1964).
206. SIMPSON, D. R.: Carbonate in hydroxylapatite. Science **147**, 501—502 (1965).
207. SIMPSON, D. R.: Substitutions in apatite: I. Potassium-bearing apatite. Amer. Mineral. **53**, 432—444 (1968).
207a. SIMPSON, D. R.: Substitution in apatite: II. Low temperature fluoride-hydroxyl apatite. Amer. Mineral. **53**, 1953—1964 (1968).
208. SIMPSON, D. R.: Partitioning of fluoride between solution and apatite. Amer. Mineral. **54**, 1711—1719 (1969).
209. SKINNER, H. C. W.: X-ray diffraction analysis techniques to monitor composition fluctuations within the mineral group: apatite. Appl. Spectrosc. **22**, 412—414 (1968).
210. STORMER, J. C., and I. S. E. CARMICHAEL: Fluorine-hydroxyl exchange in apatite and biotite: a potential igneous geothermometer. Contr. Mineral. Petrol. **31**, 121—131 (1971).
211. STOW, S. H.: The occurrence of arsenic and the color-causing components of Florida land-pebble phosphate rock. Econ. Geol. **64**, 667—671 (1969).
212. STRUNZ, H.: Mineralogische Tabellen, 5. Auflage. Leipzig: Geest & Portig KG., 1970, 621 pages.
213. SWAINE, D. J.: The Trace-Element Content of Fertilizers. Tech. Comm. **52**, Farnham Royal, England: Commonwealth Agr. Bureaux, 1962, 306 pages.
213a. TABORSZKY, F. K.: Chemismus und Optik der Apatite. Neues Jahrb., Mineral. Monatsh. **1972**, 79—91 (1972).
214. TOOMS, J. S., C. P. SUMMERHAYES, and D. S. CRONAN: Geochemistry of marine phosphate and manganese deposits. Oceanogr. Mar. Biol. Ann. Rev. **7**, 49—100 (1969).
215. TRAUTZ, O. R.: Crystallographic studies of calcium carbonate phosphate. Ann. New York Acad. Sci. **85**, 145—160 (1960).
216. TRAUTZ, O. R., R. ZAPANTA-LEGEROS, and J. P. LEGEROS: Effects of magnesium on various calcium phosphates. II Jour. Dental Res. **43**, 751 (1964).
217. TRÖMEL, G., und W. EITEL: Die Synthese von Silikatapatiten der Britholith-Abukumalit-Gruppe. Zeits. Kristallogr. **109**, 231—239 (1957).
218. TROTTER, J., and W. H. BARNES: The structure of vanadinite. Canadian Mineral. **6**, 161—173 (1958).
219. TRUEMAN, N. A.: A Petrological Study of Some Sedimentary Phosphorite Deposits. Bull. Australian Mineral Development Labs. No. **11**, 1—71 (1971).
220. TUREKIAN, K. K.: Oceans. Englewood Cliffs, N. J.: Prentice-Hall, Inc., 1968, 120 pages.
221. URIST, M. R.: Origins of current ideas about calcification. Clinical Orthopaedics **44**, 13—39 (1966).
222. VALLEE, B. L.: Biochemistry, physiology and pathology of zinc. Physiol. Rev. **39**, 443—490 (1959).

223. VAN WAZER, J. R.: Phosphorus and Its Compounds, Vol. I. New York: Interscience Publishers, 1958, 954 pages.
224. VASIL'EVA, Z. V.: [On the role of manganese in apatites.] Mem. All-Union Mineral. Soc. **87**, 455–468 (1958) [In Russian].
225. VASIL'EVA, Z. V., M. A. LITOSAREV, and N. Y. ORGANOVA: [Natural sulfate-apatite.] Dokl. Akad. Nauk SSSR **118**, 577–580 (1958).
225a. VATASSERY, G. T., W. D. ARMSTRONG, and L. SINGER: Determination of hydroxyl content of calcified tissue mineral. Calc. Tissue Res. **5**, 183–188 (1970).
226. VEITCH, F. P., and L. C. BLANKENSHIP: Carbonic anhydrase in bacteria. Nature **197**, 76–77 (1963).
227. VERNADSKY, W. I.: On the terrestrial alumophosphorous and alumosulphurous analogues of kaolinic alumosilicates. Dokl. Akad. Nauk SSSR **18**, 287–294 (1938) [Internatl. Ed.].
228. VILLIERS, J. E. DE: The carbonate-apatites; francolite from the Richtersveld, South Africa. Amer. Jour. Sci. **240**, 443–447 (1942).
229. VINCENT, J.: Microscopic aspects of mineral metabolism in bone tissue with special reference to calcium, lead and zinc. Clinical Orthopeadics **26**, 161–175 (1963).
230. VOLBORTH, A.: Phosphatminerale aus dem Lithiumpegmatit von Viitaniemi, Eräjärvi, Zentral-Finnland. Ann. Acad. Sci. Fennicae [ser. A] **3**, No. 39, 90 pages (1954).
231. VOLODCHENKOVA, A. I., and B. N. MELENT'EV: Apatites of two textural types from apatite-nepheline rocks of Chibiny. Dokl. Akad. Nauk SSSR **39**, 34–35 (1943) [In English].
232. VOROB'EVA, O. A.: Editor of Apatity. Moscow: Izd. "Nauka", 1969.
233. WACHTEL, A.: The effect of impurities on the plaque brightness of a 3000° K calcium halophosphate phosphor. Jour. Electrochem. Soc. **105**, 256–260 (1958).
234. WALLAEYS, R.: Étude d'une apatite carbonatée obtenue par synthèsis dans l'état solide. C. R. Colloque Union Internat. Chim. pure et appl., Münster, p. 183–190 (1954).
235. WALTERS, L. J., JR., and W. C. LUTH: Unit-cell dimensions, optical properties, halogen concentrations in several natural apatites. Amer. Mineral. **54**, 156–162 (1969).
235a. WANMAKER, W. L., J. W. TER VRUGT, and J. G. VERLIJSDONK: Synthesis of new compounds with apatite structure. Philips Res. Repts. **26**, 373–381 (1971).
236. WAPPLER, G.: Dielektrische Messungen an Einkristallen von Mineralien. Zeits. phys. Chem. **228**, 33–38 (1965).
237. WARING, C. L., and N. CONKLIN: Quantitative spectrochemical determination of minor elements in apatite. U.S. Geol. Surv., Prof. Paper **550-C**, 228–230 (1966).
238. WATSON, T. L., and S. TABER: Geology of the Titanium and Apatite Deposits of Virginia. Virginia Geol. Surv., Bull. **3 A**, (1913) 308 pages.
239. WELLS, R. C.: Analyses of Rocks and Minerals. U.S. Geol. Surv., Bull. **878** (1937) 134 pages.
240. WHIPPO, R. E., and B. L. MUROWCHICK: The crystal chemistry of sedimentary apatites. Trans. Amer. Inst. Mining Metal. Engrs. **238**, 257–263 (1967).
241. WILLARD, H. H., and O. B. WINTER: Volumetric method for determination of fluorine. Ind. & Eng. Chem., Anal. Ed. **5**, 7–10 (1933).
242. WINCHELL, A. N., and H. WINCHELL: The Microscopic Characters of Artificial Inorganic Solid Substances. New York: Academic Press, 1964, 439 pages.
243. WINTER, H.: Versuche zur Bildung von Apatiten und wagneritähnlichen Verbindungen des MgO, BaO und SrO. Inaug. Dissertation, Leipzig, 1913.
244. WOLLENTIN, R. W.: Lead and manganese-activated cadmium fluorophosphate phosphors. Jour. Electrochem. Soc. **103**, 17–23 (1956).
245. WONDRATSCHEK, H.: Untersuchungen zur Kristallchemie der Blei-Apatit (Pyromorphite). Neues Jahrb. Mineral. Abh. **99**, 113–160 (1963).

246. Wyckoff, R. W. G., and A. R. Doberenz: The strontium content of fossil teeth and bones. Geochim. Cosmochim. Acta **32**, 109–115 (1968).
247. Wyllie, P. J.: Experimental studies of carbonatite problems: the origin and differentation of carbonatite magmas. In: Carbonatites, (O. F. Tuttle and J. Gittins, ed.). New York: Interscience Publishers, 1966, pates 385–413.
248. Yefimov, A. F., S. M. Kravchenko, and Z. V. Vasil'eva: [Strontiapatite — a new mineral.] Dokl. Akad. Nauk **142**, 439–442 (1962) [transl. **142**, 113–116].
249. Yochelson, E. L.: Biostratigraphy of the Phosphoria, Park City, and Shedhorn Formations. U.S. Geol. Surv., Prof. Paper **313-D**, 571–660 (1968).
250. Yoon, H. S., and R. E. Newnham: Elastic properties of fluorapatite. Amer. Mineral. **54**, 1193–1197 (1969).
251. Young, E. J., and E. L. Munson: Fluor-chlor-oxy-apatite and sphene from Crystal Lode pegmatite near Eagle, Colorado. Amer. Mineral. **51**, 1476–1493 (1966).
252. Young, E. J., A. T. Myers, E. L. Munson, and N. M. Conklin: Mineralogy and geochemistry of fluorapatite from Cerro de Mercado, Durango, Mexico. U.S. Geol. Surv., Prof. Paper **650-D**, 84–93 (1969).
253. Young, R. A., and J. C. Elliott: Atomic-scale bases for several properties of apatites. Arch. oral Biol. **11**, 699–707 (1966) [cf. Trans N.Y. Acad. Sci. II, **29**, 949–959 (1966)].

Geographical Index and References to Localities *

Åbo, Finland Table 2.1, 5
Adirondack Mts., New York [131] 60 see also Essex Co.
Agulhas Bank, S. Africa [186a] 48
Airolo, Switzerland Fig. 7, 8
Aldan, Table 9.4, 62 see also sulfate apatite [225]
Allendorf, Saxony, Germany [25] Table 7.2, 43
Amelia Co., Virginia [177a] 61
Arkansas see Magnet Cove
Asio, Japan [35] 16

Baja California, Mexico [42a] 48
Bamle see Ödegården
Beaufort Co., North Carolina [197a] 48
Black Hills see morinite [61a, 62]
Bluffton, Ohio [145, 153] Table 7.2, 43, 70, 73, 88
Bob's Lake, Ont., Canada [97] 19, 87
Buckfield, Maine [138] 60
Burpala intrusive, North Baikal region [190] Table 9.4, 61, 62
Busumbu, Uganda [43] 62, 63

California see Crestmore and Santa Clara Co.
Caracoles, Chile [200] see caracolite
Cavorgia (Tavetsch) Fig. 4, 7
Cerro de Mercado, Durango, Mexico [67, 170, 252] 8, 9, 11, 15, 16, 54, 64, 65
Ceylon see Matale
Chalupki, Poland [182] 64
Chárgáon, Nagpur, India [90] 6
Chile see Caracoles
Chupa Karelia Table 9.4, 62
Coal Measures [45] 15, 89
Colorado 58 see also Eagle

Cornwall see St. Michaels Mt.
Crestmore, Riverside Co., California [50, 137] 65 see also ellestadite and wilkeite

Dehrn, Germany [122] see dehrnite
Derbyshire, Gr. Brit. [187] 52
Devonshire Table 7.2, 52 see also Tavistock
Durango see Cerro de Mercado and Peñascos de la Industria

Eagle, Colorado [251] 61
Ehrenberg near Ilmenau, Germany 56
Eplény, Hungary [76] 66
Eräjärvi, Finland [230] 60
Essex Co., New York [131, 167] 60, 66
Estremadura, Spain [89] 39

Fairfield, Utah [123] see dehrnite and lewistonite
Faraday Township, Ont., Canada [82a] 62
Florida [39, 211] 54
François Lake, B. C., Canada [189] 64

Georgia see Holly Springs
Gifu Prefecture, Japan [183] 25
Guli Table 9.4, 62

Hebron, Maine [185a] Fig. 4, 7
Hitterö, Norway [98] 58
Holly Springs, Cherokee Co., Georgia [177] 65
Hospenthal see Kemmelten

Idaho see Phosphoria Formation
Illinois [73] Table 7.1, 41
Ilmenau, Germany 56
Il'meny Table 9.4, 62
India see Chárgáon

* The name of a locality may not appear on the page indicated, but may be identified by the reference cited. A few names are included of sedimentary formations, intrusive masses, pegmatites, etc.

Israel [173 a] 52

Jiwaara, Kuusamo, Finland 56
Jordan [159, 195] 55

Kangerdluarsuk, Greenland [90] Table 2.2, 6
Katzenbuckel, Odenwald, Germany 64
Kazakhstan Table 9.4, 62
Kemmelton, Switzerland [177] 66
Kenya, Africa Table 9.5, 62
Keystone, South Dakota [106] 15
Khibiny, Kola Peninsula, U.S.S.R. [48, 110, 231] Tables 9.3 and 9.4, 1, 58, 59, 60, 62, 64
Kiriaebinsk, Miask Fig. 1, 7
Kivu, Congo Table 9.5, 63
Kola Peninsula see Khibiny
Kovdor Table 9.4, 62
Kragero, Norway Table 9.4, 62
Kusa, Urals, U.S.S.R. Table 9.4, 62
Kushva, Urals, U.S.S.R. Table 9.4, 62
Kutná Hora, Czechoslovakia [95] Table 7.2, 43, 64, 89
Kuusamo see Jiwaara

Lausitz, Germany 56
Libby, Montana [124] 59
Lovozero Table 9.4, 62

Magnet Cove, Arkansas [32, 163] Fig. 13 Table 7.3, 1, 15, 21, 44, 64
Maine see Buckfield and Hebron
Mama, E. Siberia Table 9.4, 62
Matale, Ceylon [40] 30
Mbeya, Tanzania Table 9.5, 63
Milburn, Otago, New Zealand [100] 64
Mineville district, Essex Co., New York [131, 167] 60, 66
Mittelgebirge, Bohemia [Czechoslovakia] 56
Montana see Libby and Phosphoria Formation
Moon [3, 71, 129] 66

Naegi, Gifu Prefecture, Japan [183] 25
Nassau see Dehrn
Nelson Co., Virginia [238] 1, 58
New Jersey see Sussex Co.
New York see Adirondack Mts. and Mineville district
North Carolina see Beaufort Co.
Nuevo León see Sabinas Hidalgo

Odenwald 56, 64
Ödegården, Bamle, Norway [23, 179] 64
Ohio see Bluffton
Oka, Quebec, Canada [72 a] 62
Oldonyo Dili, Tanzania Table 9.5, 63
Ontario see Bob's Lake and Faraday Township
Otago see Milburn

Peñascos de la Industria, Durango, Mexico [67] 64, 65
Pfitsch, Tyrol, Austria Figs. 2 and 3
Phosphoria Formation [4, 5, 29, 82, 87, 128, 203, 219, 240, 249] Table 8.1, 27, 48, 51
Pisek, Bohemia [Czechoslovakia] 9
Polar Urals, U.S.S.R. [27] Table 9.4, 62

Quebec see Oka

Rangwa, Kenya Table 9.5, 63
Richtersveld, Cape Province, S. Africa [228] 64
Riverside see Crestmore
Robin Hood Quarry, Yorkshire, Gr. Brit. [45] 15, 89
Rongstock, Bohemia [Czechoslovakia] 56
Rossberg, Odenwald, Germany 56
Ryukyu Islands, Pacific Ocean [108] 52, 53

Sabinas Hidalgo, N. L., Mexico [30, 152] 46, 66
St. Michaels Mt., Cornwall, Gr. Brit. Fig. 5, 7
St. Paul's Rocks, Atlantic Ocean 64
Saitama Prefecture, Japan [86] Table 9.6, 30, 65
Santa Clara Co., California [197] 6
Schlaggenwald, Czechoslovakia Fig. 6, 8
Schwaden, Bohemia [Czechoslovakia] 56
Shabry, Urals Table 9.4, 62
Slyudyanka Table 9.4, 62
South Dakota see Keystone and morinite
Staffel, Germany [25, 80] Table 7.2, 6, 27, 43, 45, 64
Sussex Co., New Jersey [104] see spencite
Sweden [72] 60, 65 see also Varuträsk

Tanganyika, Africa Table 9.5, 63
Tanzania, Africa Table 9.5, 63
Tavistock, Devon., Gr. Brit. [198] Table 7.2, 43
Thorpe-on-the-Hill see Robin Hood Quarry
Tonopah, Nevada 64

Transbaikal Table 9.4, 62
Tyrol *see* Pfitsch

Uganda *see* Busumbu
Urals, Table 9.4, 62
Úrkút, Hungary [76] 66
Utah *see* Phosphoria Formation and Fairfield
Uzunosawa, Saitama Prefecture, Japan [86] 30

Val Sessera, Italy [188] 58
Varuträsk [pegmatite], Sweden [194] 60
Viitaniemi *see* Eräjärvi

Virginia *see* Amelia Co. and Nelson Co.
Vishnevyye Mts. Table 9.4, 62
Vuori Yarvi Table 9.4, 62

Western district [Idaho, Montana, Utah and Wyoming] *see* Phosphoria Formation
Wyoming *see* Phosphoria Formation

Yakutia [248] 61, 62, 90
Yorkshire, Gr. Brit. [187] 52, *see also* Robin Hood Quarry

Zambia, Africa Table 9.5, 63

Subject Index*

Absorption *also see* infared
— theories 4, 44, 52, 85
Abukumalite 25, 26, 32, 61
Abundance [trace elements] 51
Accuracy 83
Achrematite 26
Activators 33
Age determination 48, 53, 65
Algae 78
Alkalies *see* potassium and sodium
Alkaline phosphatase 76
Aluminum 3, 25, 32, 33, 34, 54, 61
Ammonium citrate 36, 54
"Amorphous" 44, 71, 72, 75, 82, 86
Analysis 4, 36, 41, 83
"Anorganic" bone 84
Antimony 33
Apatite-nepheline 60
Apatite-villiaumite 60
Arsenic 20, 54
Arthropods 79
Assimilation 68
Authigenic minerals 50
Axial ratios 7

Bacteria 77, 86
Barium 25, 33, 34, 54, 78
Bastnaesite 50, 60
Beckelite 26
Bellite 26
Belovite 25
Biologic influences 49
Birefringence 5, 9, 44, 45, 57
Bismuth 34
Blood 68, 74
Blue color 15
Bone 53, 69, 71, 76, 77
Bone [analysis] 69, 72, *also see* fossil bone
"Bone char" 30, 38, 49

"Bone salt" 68
Boron 29, 35, 53
Brachiopod shells 49, *also see Lingula*
Brightness 33
Britholite 26, 32, 61
Bromine 29, 35
Brushite 10, 37, 49, 71, 74, 80, **87**

Cadmium 23, 24, 33, 34
"Calcium-deficient apatites" 71
Calcium phosphate 36, 38, 76, 84
Ca/Mg ratios 49
Ca/P ratios *see* ratios
Caracolite 32
Carbon dioxide [determination of] 83
Carbonate content 36, 54
— groups 5, 19, 21, 22, 30, 31, 40, 44
Carbonatites 63
Carbonic anhydrase 76
Cardiovascular mineralization 79
Caries *see* dental caries
Catalysts 49, 79
Cerium 25, 29, 33, 34, 54, 61
"Channels" [in structure] 19, 30
Chemical compound [vs. mineral] 2
— methods of analysis 4, 83
— weathering *see* weathering
Chlorapatite 19, 21, 28, 47, **87**
Chloride ions [in sea water] 50
Chlorspodiosite 37
Chromium 28, 35, 54, 82
Citric acid 69
Classification [of rock phosphates] 48
Clay minerals 50, 54
Cleavage 9, 16
Cloudiness [of crystals] *see* inclusions
Cobalt 15, 23, 54, 78
Coefficients [dimensional] 42
Colloid science 4

---

* Page numbers in boldface indicate data given in the Appendix.

Collophane 10
Color [of luminescence] 33
Coloration 9, 15, 26, 57
Compositional model 12
Compressibility 16
Conodonts 79
Contact deposits see metamorphic rocks
Contamination 27, 36, 50, 60, 67, also see inclusions
CO_3OH substitution 45
Coordination numbers 23
Copper 23, 33, 53, 54, 78
Corneal calcification 79
Crabs 79
Crystallinity 10, 44, 72

Dahllite 10, 26, 41, 43, 49, 64, 68, 70, 73, 77, **88**
Defluoridization of water 38, 49
"Degree of crystallinity" 44, 72
Defects 43
Deficiency [of P] 41, 46
Dehrnite 10, 23
Density 9, 11, 16, 46, 57
Dental calculus 70, 80
— caries 38, 75, 77
— enamel 68, 69, 70, 72, 86
Dentifrices 33, 38, 80
Dentin 73
Depletion [of Ce] 54
Detergents 1
Detrital minerals 50
Deuterons 29, 42
Diadochy 32
Dielectric constant 16
Differential thermal analysis 39
Diffraction method 73, 81
Diffuse scattering 73
Dikes 60
Dimensions [unit cell] 11, 24, 42, 44
Direct precipitation 49
Discrimination [Sr-90] 77
Dislocations see "screw dislocations"
Dispersion 15
"Double salts" 3, 76
Dynamics [of bone formation] 80

Economic utilization 1, 48, 50
Elastic properties 16
Electric currents 79
Electromotive force 79
Electron capture 82

Electron diffraction 73, 75
— probe analysis 83
— spin resonance spectra 82
Electrostatic charges [balancing of] 22
Ellestadite 2, 3, 22, 26, 31, 65, **88**
Endlichite 26
Energy relations 39, 51, 60
— transfer 34
Enrichment 51, 52, 53, 55
Enzymes 76, 77
Epitaxy 74
Equilibrium 37, 85
Errors [analytical] 83
Ethylenediamine 75, 84
Ettringite 40
Eupyrchroite 10
Europium 23, 25, 34
Eutectic 37, 60
"Excess" water 66, 70
Extinction [optical] 15

Factor-group 21
Fermorite 10, 26
Fertilizers 1, 52, 53
"Fibrous francolites" 44
Fish [incl. scales] 53, 79
Fission tracks 65
Fluorapatite **88**
Fluorescence 15, 61
Fluorine [marine geochemistry of] 50 [removal from water supplies] see "bone char" [analysis for] 4 [in bone] 78
Fluorosis 75
Forms 8, 9
Fossil bones 38, 53, 70
Fossilization process 38, 53
Francolite 6, 10, 22, 26, 27, 41, 43, 44, 45, 49, 64, 68, **88**
Frequency [of forms] 8
— changes [IR] 21

Gadolinium 62, 66
Gar-pike [scales] 79
Geochemistry [sedimentary] 50, 54
Germanium 28
GLADSTONE & DALE 13
Glass 35, 72, 82
Goniometry 6, 9
Granite 56
Green color 15, 27
Gypsum 50

Subject Index

Habit 7, 21
Halogen atoms 24, 29
Halophosphates 33, 35
Hard tissues 1, 68
Hardness 9, 16
Hematite *see* iron ores
Hexagonal symmetry 19, 24, 29
Histoplasmosis 79
H_3O^+ 26, 40, 44, 55
H_4O 29, 41, 42, 44, 55
Hyaline cartilage 79
"Hydrated carbonate apatites" 41, 69
Hydrothermal apatite 64
Hydroxyapatite 17, **89**
Hydroxylellestadite 46, 65, **88**
Hypothetical phases 39

Igneous rocks 56
"Ignition loss" 4, 45, 84
Ilmenite *see* iron ores
Inclusions 11, 58
Infrared 4, 21, 35, 44, 67, 73, 81, 82, 84
Inhibitors 49, 76, 77
Intensities [*vs* frequencies] 9
    [x-ray diffraction] 17
Invertebrates 68, 86
Iodine 29, 35, 78
"Ion exchange" 85
Ionic radii 22, 28, 52
— refractivities 13, 45
— strength 49
Iron [in bone] 78
— [in phosphorites] 54
— [substitution of] 5
— ore [association with] 1, 60, 64, 65
Isomorphic substitution 3, 22, 39
Isomorphism 2
Isotopes 55, 76, 77
Isotypes 11, 21
Isotypism 2
Isular deposits 48, 53

Junction potentials 79

KARL FISCHER titration method 84
Kurskite 10

Lanthanum 15, 33, 34, 58
Leaching 36, 54, 70
Lead 16, 23, 33, 34, 54
Lessingite 26
Lewistonite 10, 23

"Line splitting" 19, *also see* splitting
Lingula 49, 73, 79
Liquidus 37, 60
Localities *see* Geographical Index
LORENTZ & LORENZ 13
Loss on ignition 4, 45, 84
Luminescence 33, 61
Lunar apatites 66
Luster 9

Magnesium 23, 34, 49, 77, 78
Magnetic susceptibility 16
Magnetite *see* iron ores
Manganese 5, 6, 15, 23, 24, 32, 33, 34, 60, 62, 78
MnO_4^{-3} 32, 35
Mechanical properties 16
Melting points 16
Mercury 53
Merrillite 66
Metabolic processes 78
Metamict 60
Metamorphic rocks 65
Metasomatized facies 58, 60
Meteoritic apatites 66
Mimetite 21, 26
Mineral [*vs.* chemical compound] 2
Mineralization [biologic] 76, 80, 85
Mining *see* production
Mirror symmetry 18
Mixtures 36, 47
Molar refractivity *see* ionic refractivity
Molecular weight 46
Mollusks 79
Molybdinum 53, 78
Monazite 50, 58, 60
Monetite 49, 80, **89**
Monoclinic symmetry 19, 21
Morinite 25, 32, 61, 82, **89**
Moroxite 10
Morphology 8, 72
Multiplicity factors 8

Nauruite 10
Neisseria see bacteria
Neodymium 15, 34
Nepheline-apatite 60
Neutron activation 78
— diffraction 42, 82
Nickel 15, 23, 34, 54
Nitrogen 69
Nodules 48

Normative minerals 56, 57
Nuclear magnetic resonance 70, 82

Octacalcium phosphate 36, 37, 80, **91**
OH_3 ion *see* H_3O^+
Openings *see* "channels"
Optical properties 15, 46, *also see* refractive index
Oral calculus 77
Organisms 49
Osteoblasts 76
Osteoid tissue 80
Oxyapatite *see* voelckerite

Paragenetic relations 60, 65
Paramagnetic centers 82
Parameters [structural] 18
Pathological calcification 79, 86
PAULING's principle 19
Pegmatites 60, 61
PENFIELD method 41, 84
Peridotite 58
Petrologic paragenesis 60, 65, 85
Phases [hypothetical] 39
Phosphorescence 61
Phosphoria Formation *see* Geographical Index
Phosphorite 10, 29, 33, 39, 48
Phosphors 33
Phosphorus [recycling] 68
Piezoelectric properties 16, 21, 79, 83
Planar groups 44
Plants 68
Pleochroism 15
Pneumatolytic 64
Podolite 10
Polarized radiation [IR] 82
Polymorphism 2, 12, 75
Polyphosphates 77
Polysynthetic twinning 15, 21
Portlandite 37
Potassium 23, 77, *also see* lewistonite
Praseodymium 15, 29
Precision 83
Precursors 71, 74
Prismatic *see* habit
Production 1
Protons 42
Pseudosymmetry 19
Pyroelectric properties 21, 83
Pyromorphite 26, 29, 30
Pyrophosphate 77
Pyroxene 58

Quercyite 10

RAMAN spectra 21
Rare earths 16, 33, 50, 57, 58, 61, 62, 66
Rarer elements 51, 60
Ratios [Ca/P] 40, 41, 46, 69
Recycling 68
Refractive index 5, 12, 14, 36, 44, 57
Regeneration [bone] 80
Renal stone 79
Replacement 49
Resolution 73
Resorption 79, 80
Retardation *see* birefringence
Rutile *see* iron ores

Saamite 25, 26, 59
Salinity 49
Saliva 76
Sampling methods 83
Saturation [sea water] 49
Scales [fish] 79
Scandium 50
Scawtite 40
Scouring agent 38
"Screw dislocations" 21
Sea water 48, 50, 51, 68, 86
Secondary enrichment 52
Sedimentation 48, 54
Segregations 59
Selenium 53
Shellfish 77
Shrimps 79
Silicate apatites 65
Silicon 25, 29, 31, 34, 54
Sills 60
Single crystals [$carHap$ or $carFap$] 21, 81
Site-group 21
Skeletal tissues *see* bone
Sodium 23, 32, 34, 77, *also see* dehrnite
— tripolyphosphate 1
Solubility product 50, 68, 85
Space group 17
Specific gravity *see* density
Spencite 29
Splitting [IR bands] 21
Spodiosite 37
Stability 22
Staffelite 10
Standartization 83
Statistical vacancy 29, 30, 34
Stoichiometric relations 2, 36, 40, 42, 82

Strontiapatite 26, 61, 90
Strontium 14, 16, 23, 25, 33, 34, 53, 60, 61, 78
Strontium-90 77
Sulfanilamide 76
Sulfate apatite 26, 65
— ions [in sea water] 50
Sulfate-silicate apatites 31, 65
Sulfur 25, 29, 31
Superphosphate 55
Superstructure 19
Surface area 44, 70, 85
"Surface chemistry" 75, 85
Svabite 10, 26
Symmetry 17
Synonyms 10
Synthesis 23
Synthetic apatites 34

Tabular *see* habit
Teeth 53, 68, 69, 71, 76, 77, *also see* dental enamel
Tetrahedral groups 29, 31
Thallium 34
Thermogravimetric analysis 39
Thermoluminescence 61
Thorium 25, 29, 50, 58
Tin 23, 25, 33, 34
Titanium 54, 57
Toothpastes *see* dentifrices
Toxicity [of Be] 33 [of Se] 53
Trace elements 51, 53, 78
Triangular groups 31, 44
Tricalcium phosphate *see* calcium phosphate
"Tricalcium phosphate hydrate" 37, 76
Triclinic symmetry 21, 81
Tungsten 28

"Tunnels" *see* "channels"
Twinning 10, 15, 21
Twins 10, 81
"Type A carbonate apatite" 31

Ultrabasic rocks 56, 58
Uranium 29, 50, 52, 58
Urinary "stones" 79
Utilization *see* economic utilization

Vacancies [structural] 29, 30, 34
Vanadinite 11, 12, 21, 26, 28
Vanadium 27, 50, *also see* vanadinite
Varietal names 10
Vegard's law 42
Veins 60
Vibrational spectra 21
Villiaumite-apatite 60
Voelckerite 6, 10, 19, 35, 41, 84, **91**
Volume [unit cell] 11, 14

Water [determination of] 4, 41, 45, 84
 [chemically bound] 40, 70, 77
 [supplies] 49
Weathering 53, 85
Whitlockite 19, 37, 49, 66, 80, **90**
Wilkeite 10, 22, 26, 31, 65, **90**
Wood tissue 10, 49

Xenotine 58

Yttrium 25, 29, 34, 50, 61, 62

Zeolitic characteristics 30
Zinc 23, 34, 53, 77, 78
Zirconium 33

Printer: R. Spies & Co., A-1050 Wien